# Atomic structure

## Test your knowledge

1. a) Solids are dense because their particles are _____ .

   b) Particles in liquids are able to move, therefore liquids can _____ .

   c) Gases are easily compressed because their particles are _____ .

2. Gases move by a process called _____ . Hydrogen diffuses quicker than nitrogen because it is _____ .

3. Atomic nuclei contain _____ and _____ . Surrounding the nuclei are _____ contained in _____ .

4. The number of protons in a nucleus is the _____ number. The total number of protons and neutrons is called the _____ number.

5. Atoms of the same element with different numbers of neutrons are called _____ .

6. A $^{27}_{13}$Al atom contains 13 protons and _____ neutrons.

7. A $^{27}_{13}$Al atom contains _____ electrons. Its electronic structure is _____ .

## Answers

**1** a) closely packed b) flow or change shape c) far apart **2** diffusion / lighter **3** protons / neutrons / electrons / shells **4** atomic / mass **5** isotopes **6** 14 **7** 13 / 2,8,3

*If you got them all right, skip to page 4*

# Atomic structure

## Improve your knowledge

**1** The arrangement of particles in matter determines their physical properties.

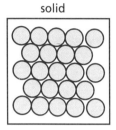
solid

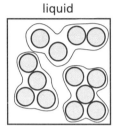
liquid

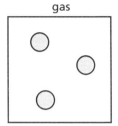
gas

| State | Arrangement of particles | Physical property |
|---|---|---|
| Solid | a) Very close together<br>b) Strong forces, particles cannot move | a) Dense and non-compressible<br>b) Fixed volume and shape |
| Liquid | a) Groups of closely packed particles that move about | a) Shape changes to fit container |
| Gas | a) Particles far apart | a) Compressible |

**2** Gases move by **diffusion**. Gas particles collide with each other and the container, changing direction.

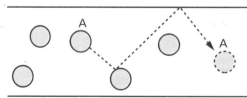

particles move by 'bouncing' off each other and the container

**3** Atoms have a small, central nucleus containing **protons** and (except hydrogen) **neutrons**. Around the nucleus are **electrons** in **shells**.

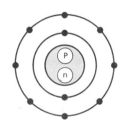

**4** The table shows the properties of protons, neutrons and electrons.

|  | Relative mass | Relative charge |
|---|---|---|
| Proton | 1 | +1 |
| Neutron | 1 | 0 |
| Electron | 0 | −1 |

**5** Atomic number is the number of protons in the nucleus. Mass number is the sum of the number of protons and neutrons in the nucleus. Relative atomic mass is the average mass of an atom compared to $1/12$ the mass of an atom of carbon-12.

**6** Isotopes are atoms of the same element with different numbers of neutrons.

mass number ⟶ $^{12}_{6}C$ or $^{13}_{6}C$ ⟵ **different** mass number
atomic number ⟶                   ⟵ **same** atomic number

**7** Electrons are located in shells. The first shell holds up to 2 electrons, all other shells up to 8 electrons. The number of electrons is the same as the atomic number. From this we can work out the electronic structure, e.g.

| Element | Atomic number | Number of electrons | Electronic structure |
|---|---|---|---|
| $_8O$ | 8 | 8 | 2,6 (2+6 =8) |
| $_{11}Na$ | 11 | 11 | 2,8,1 (2+8+1=11) |

# Check list

Are you sure that you understand these key ideas?

the structures of solids, liquids and gases / diffusion / atomic structure / atomic number / mass number / relative atomic mass / isotopes / electronic structures

*Now learn how to use this knowledge*

# Atomic structure

## Use your knowledge

20 minutes

**1** A piece of cotton wool wetted with ammonia solution and another wetted with hydrogen chloride solution were inserted into each end of a glass tube as shown. Eventually a white vapour was observed at point A.

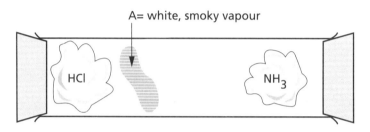

A= white, smoky vapour

a) How did the ammonia and the hydrogen chloride move along the tube?

_Hint 1_

b) How does this occur?

_Hint 2_

c) Why didn't the vapour form immediately?

_Hint 3_

d) Why did the vapour form nearer to the hydrogen chloride rather than the ammonia?

_Hint 4_

# GCSE in a week

## Chemistry

Dan Evans and Alex Watts,
Abbey Tutorial College
Series Editor: Kevin Byrne

*Where to find the information you need*

| | |
|---|---|
| Atomic structure | 1 |
| Chemical reactions and formulae | 7 |
| Bonding | 13 |
| The periodic table | 19 |
| Metals and the reactivity series | 25 |
| Non-metals | 31 |
| Rates of reactions | 37 |
| Gases and ions | 43 |
| Water | 49 |
| Salts - preparation and solubility | 55 |
| Industrial processes 1 | 61 |
| Industrial processes 2 | 67 |
| Products from ores | 73 |
| Products from oil | 79 |
| Rocks and plate tectonics | 85 |
| *Mock exam* | 90 |
| *Answers* | 94 |

Letts Educational
Aldine Place
London W12 8AW
Tel: 0181 740 2266
Fax: 0181 743 8451
e-mail: mail@lettsed.co.uk
website: http://www.lettsed.co.uk

Every effort has been made to trace copyright holders and obtain their permission for the use of copyright material. The authors and publishers will gladly receive information enabling them to rectify any error or omission in subsequent editions.

First published 1998
Reprinted 1998, 1999 (three times)

Text © Dan Evans and Alex Watts 1998
Design and illustration © (Letts Educational) Ltd 1998

All our Rights Reserved. No part of this publication may be reproduced, stored in a retrieval system, or transmitted, in any form or by any means, electronic, mechanical, photocopying, recording or otherwise, without the prior permission of Letts Educational.

**British Library Cataloguing in Publication Data**
A CIP record for this book is available from the British Library.

ISBN 1 85758 6964

Editorial, design and production by Hart McLeod, Cambridge

Printed in Great Britain by Ashford Colour Press

Letts Educational is the trading name of Letts Educational Ltd, a division of Granada Learning Ltd. Part of the Granada Media Group.

**2** Consider the unknown atoms represented below:

$$^{12}_{6}V \quad ^{14}_{6}W \quad ^{14}_{7}X \quad ^{11}_{5}Y$$

a) Which has the most protons? _____ *Hint 5*

b) Which has the most neutrons? _____ *Hint 6*

c) Which has the most electrons? _____ *Hint 7*

d) Which are isotopes of the same element? _____ *Hint 8*

**3** Carbon has a relative atomic mass of 12 and magnesium a relative atomic mass of 24. What does this tell you about the mass of one atom of magnesium compared to one atom of carbon? *Hint 9*

_____

**4** a) Suggest why air is used to pump up bicycle tyres. *Hint 10*

_____

_____

b) In terms of the arrangement of particles, explain why gases exhibit the property mentioned in (a) but liquids do not. *Hint 11*

_____

_____

**5** An unknown atom Z has a mass number of 31 and 16 neutrons.

a) Name or give the true symbol for the element. _____ *Hint 12*

b) Write its electronic structure. _____ *Hint 13*

*Hints and answers follow*

# Atomic structure

## Hints

*1* Hydrogen chloride and ammonia solutions easily turn into vapours or gases.

*2* What properties of gases allow the above process to occur?

*3* The hydrogen chloride vapour and ammonia vapour must meet before the white vapour can form.

*4* Which gas moved faster and why?

*5* Which are the atomic numbers?

*6* How can you find the number of neutrons from the mass and atomic numbers?

*7* The number of electrons is related to the number of protons.

*8* Isotopes have the same number of one type of particle.

*9* Relative atomic mass allows us to compare the masses of atoms.

*10* What happens to a tyre when it goes over a bump in the road?

*11* How are gas particles arranged differently to those in a liquid?

*12* Use your periodic table.

*13* How many electrons are there?

### Answers

**1** a) diffusion b) particles move by colliding with other particles and the walls of the tube c) it takes time for the vapours to diffuse to point A d) hydrogen chloride particles are heavier than ammonia particles and therefore diffuse more slowly **2** a) X b) W c) X d) V & W **3** twice as heavy **4** a) it is compressible and so it acts as a cushion b) its particles are far apart **5** a) phosphorus or P b) 2,8,5

# Chemical reactions and formulae

## Test your knowledge

**10 minutes**

**1** Hydrogen chloride has a relative formula mass of _____.
Calcium chloride has a relative formula mass of _____.

**2** Write the formula of the following compounds:
sodium oxide _____ calcium iodide _____

**3** Complete the following word equations:
a) zinc + _____ → zinc sulphate + copper
b) potassium hydroxide + hydrochloric acid → _____ + water

**4** Write the following word equation in symbols:
sodium + water → sodium hydroxide + hydrogen
_____

**5** Balance the following equation:
_____ $C_2H_6$ + _____ $O_2$ → _____ $CO_2$ + _____ $H_2O$

**6** Fill in the state symbols for the equation:
$CaCO_3(s)$ + $H_2SO_4(aq)$ → $CaSO_4$ _____ + $CO_2$ _____ + $H_2O$ _____

## Answers

**1** a) 36.5 b) 111  **2** $Na_2O$/$CaI_2$  **3** a) copper sulphate b) potassium chloride  **4** $Na + H_2O → NaOH + H_2$  **5** $1C_2H_6 + 3½O_2 → 2CO_2 + 3H_2O$  **6** $CaSO_4(aq)$ / $CO_2(g)$ / $H_2O(l)$

*If you got them all right, skip to page 10*

# Chemical reactions and formulae

## Improve your knowledge — 20 minutes

**1** Atoms are represented by the symbols in the **periodic table**. Molecules made up of atoms are represented using these symbols giving a **formula**.

| Compound | Contains | Formula | Rfm |
|---|---|---|---|
| sodium chloride | Na, Cl | NaCl | 23 + 35.5 = **58.5** |
| water | H, O | $H_2O$ | (2 × 1) + 16 = **18** |

The relative formula mass (rfm) can be worked out by adding together the relative atomic masses of all the atoms present as shown.

**2** **Valency** is how many bonds an atom **usually** forms, and depends on which **group** in the periodic table the element is in, i.e.

| Group | 1 | 2 | 3 | 4 | 5 | 6 | 7 |
|---|---|---|---|---|---|---|---|
| Valency | 1 | 2 | 3 | 4 | 3 | 2 | 1 |

In ionic substances the valency of each ion is equal to its **charge**.

To work out a formula write down the two symbols and their valencies and then **swap** the numbers around as shown below.

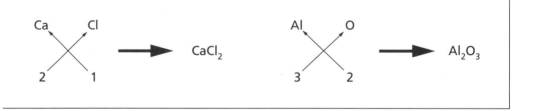

**3** Chemical reactions can be shown as **word equations.**

sodium hydroxide + hydrochloric acid → sodium chloride + water

zinc + sulphuric acid → zinc sulphate + hydrogen

calcium carbonate + sulphuric acid → calcium sulphate + carbon dioxide + water

**4** The word equations on the previous page can be written using formulae.

$$NaOH + HCl \rightarrow NaCl + H_2O$$

$$Zn + H_2SO_4 \rightarrow ZnSO_4 + H_2$$

$$CaCO_3 + H_2SO_4 \rightarrow CaSO_4 + CO_2 + H_2O$$

**5** Chemical equations must be **balanced**. **THIS MEANS THAT THE NUMBER OF EACH TYPE OF ATOM MUST BE THE SAME ON BOTH SIDES OF THE EQUATION.**

If they aren't, then you have to put numbers **in front** of the formulae containing that atom to **make** them the same, as shown below.

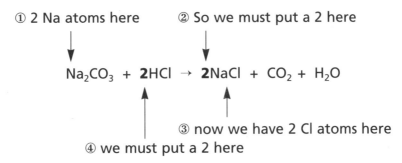

① 2 Na atoms here
② So we must put a 2 here

$$Na_2CO_3 + 2HCl \rightarrow 2NaCl + CO_2 + H_2O$$

③ now we have 2 Cl atoms here
④ we must put a 2 here

**6** The physical states of substances are shown by the following state symbols: solid - (s); liquid - (l); gas - (g); aqueous, (ions in water) - (aq).

e.g. $Na_2CO_3(s) + 2HCl(aq) \rightarrow 2NaCl(aq) + CO_2(g) + H_2O(l)$

# Check list

**Are you sure you understand these key terms?**

**symbol / formula / rfm / valency / chemical equation / word equation / balancing equations / state symbol**

*Now learn how to use this knowledge*

# Chemical reactions and formulae

## Use your knowledge

**20 minutes**

**1** Phosphorus is in group 5 and reacts with chlorine to form a compound that contains **only** phosphorus and chlorine.

a) Write down the formula of the product you would expect when phosphorus reacts with chlorine.

*Hint 1*

_____

b) Write a word equation for the formation of this compound from its elements.

*Hint 2*

_____

c) Arsenic, As, can have a valency of 5. Write down the formula of arsenic chloride.

*Hint 3*

_____

**2** Magnesium sulphate, $MgSO_4$, has an rfm of 120. Solid magnesium sulphate is often contaminated with water $MgSO_4 xH_2O$, where x represents the number of water molecules.

a) If the rfm of $MgSO_4 xH_2O$ is 246, calculate the mass of water present in the formula.

*Hint 4*

_____

b) Calculate the number of water molecules (x).

*Hint 5*

_____

**3** a) Balance the equation for propane, $C_3H_8$ burning in air. *(Hint 6)*

___ $C_3H_8$ + ___ $O_2$ → ___ $CO_2$ + ___ $H_2O$

b) If propane is burnt in a small amount of air it forms carbon monoxide, CO, instead of carbon dioxide. Balance the equation for this reaction.

___ $C_3H_8$ + ___ $O_2$ → ___ CO + ___ $H_2O$

c) Ethanol, $C_2H_5OH$, burns to form the same products as propane. Write a word equation for the reaction when ethanol burns in excess air. *(Hint 7)*

_____

**4** Aluminium reacts with oxygen to form aluminium oxide.

a) What is the formula of aluminium oxide? _____ *(Hint 8)*

b) Balance the equation for the formation of aluminium oxide.

___ Al + ___ $O_2$ → ___ Al ___ O ___ *(Hint 9)*

c) This oxide reacts with dilute sulphuric acid to produce a solution of aqueous $Al^{3+}$ ions. If aqueous sodium hydroxide is added to this solution a precipitate of aluminium hydroxide, $Al(OH)_3$ is formed. Balance the equation for this reaction, showing state symbols.

___ $Al^{3+}$ ___ + ___ $OH^-$ ___ → ___ $Al(OH)_3$ ___

**5** Lead forms two chlorides, $PbCl_4$, a liquid, and $PbCl_2$, a solid. When $PbCl_4$ is heated it forms $PbCl_2$ as one of the products. Complete the equation for this reaction, showing state symbols. *(Hint 10)*

___ $PbCl_4$ ___ → ___ $PbCl_2$ ___ + _____

*Hints and answers follow*

# Chemical reactions and formulae

1. What is the normal valency for group 5 atoms?

2. What are the reactants and products of this reaction?

3. How many chlorine atoms will arsenic bond to if it has a valency of 5?

4. What mass is due to the water present?

5. Compare the rfm of one $H_2O$ molecule with the total mass of water.

6. Count the atoms of each type on both sides of the equation.

7. What are the reactants and products?

8. Work out the formula from the charges on the ions.

9. Count the atoms of each type on both sides of the equation.

10. What must the other product be?

## Answers

1 a) $PCl_3$ b) phosphorus + chlorine → phosphorus chloride c) $AsCl_5$ 2 a) 126 b) 7 3 a) $C_3H_8$ + $5O_2$ → $3CO_2$ + $4H_2O$ b) $C_3H_8$ + $3\frac{1}{2}O_2$ → $3CO$ + $4H_2O$ c) ethanol + oxygen → carbon dioxide + water 4 a) $Al_2O_3$ b) $2Al + 1\frac{1}{2}O_2$ → $Al_2O_3$ c) $Al^{3+}(aq) + 3OH^-(aq)$ → $Al(OH)_3(s)$ 5 $PbCl_4(s)$ → $PbCl_2(s) + Cl_2(g)$

# Bonding

## Test your knowledge

**1** Atoms form enough bonds to get _____ in their outer shell.

**2** Ionic bonds form by electron _____ . When atoms lose electrons they form _____ ions called _____ . When atoms gain electrons they form _____ ions called _____ .

**3** What would the charge be on the ions formed by the following atoms?
K _____   Ca _____   Br _____ .

**4** Write down the electronic structure for the **ions** you would expect for the following atoms: Ca = _____   O = _____ .

**5** Covalent bonds are formed when two atoms _____ electrons and usually occurs between two _____ atoms.

**6** a) Ionic substances have _____ structures containing many _____ bonds **so** they have _____ melting points.

b) These structures contain ions and so often dissolve in _____ and can conduct _____ when melted.

c) Covalent **molecular** substances have _____ melting points than **giant** covalent substances.

## Answers

1 8 electrons  2 transfer / positive / cations / negative / anions  3 +1 / +2 / −1  4 2,8,8 / 2,8  5 share / non-metal  6 a) giant / strong / high  b) water / electricity  c) lower

If you got them all right, skip to page 16

# Bonding

## Improve your knowledge

**20 minutes**

**1** Atoms can combine by using electrons to form **chemical bonds**. They can form **ionic** bonds or **covalent** bonds. They usually form enough bonds to get **8 electrons** in their outer shell.

**2** Ionic bonds occur by **electron transfer** from one atom to another, forming positive ions called **cations** and negative **ions** called **anions**. The two ions are held together because positive and negative things attract each other. Sodium chloride is shown below.

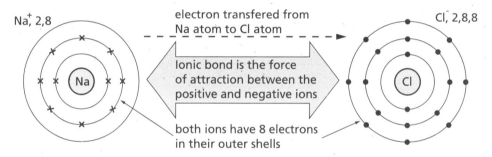

**3** Metals form cations by losing electrons, non-metals form anions by gaining electrons, in order to get 8 electrons in their outer shell.

| Group | 1 | 2 | 3 | 4 | 5 | 6 | 7 |
|---|---|---|---|---|---|---|---|
| Ion charge | +1 | +2 | +3 | None | -3 | -2 | -1 |
| Example | $Na^+$ | $Ca^{2+}$ | $Al^{3+}$ | None | $N^{3-}$ | $O^{2-}$ | $Br^-$ |

Ionic bonding usually occurs between a metal atom and a non-metal atom.
e.g.   sodium chloride, NaCl   magnesium oxide, MgO

**4** The electronic structure of ions can be worked out by adding or subtracting the electrons gained or lost from the atomic electronic structure, e.g. $Mg^{2+}$ = 2,8 and $Cl^-$ = 2,8,8

**5** Covalent bonding occurs when atoms **share** two electrons, forming enough bonds to get 8 electrons in their outer shells.

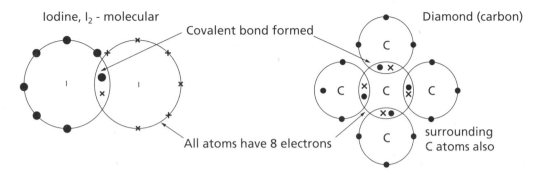

Covalent bonding usually occurs between two non-metal atoms e.g. water, $H_2O$ and chlorine, $Cl_2$.

**6** There are three main structures: **giant ionic, giant covalent** and **molecular covalent**.

| Structure | Description | Property |
|---|---|---|
| **Giant ionic** e.g. sodium choride, NaCl | i) All ions strongly bonded ii) Contains charged ions iii) Ions free to move when dissolved or melted | i) High melting point ii) Dissolves in water iii) Conducts electricity in these states |
| **Giant covalent** e.g. Diamond, C | i) All atoms strongly bonded ii) No charged ions | i) High melting point ii) Does not dissolve in water or conduct electricity (except graphite) |
| **Molecular covalent** e.g. Iodine, $I_2$ | i) Weak forces between molecules ii) No charged ions | i) Low melting point ii) Does not dissolve in water or conduct electricity |

# Check list

Are you sure you understand these key terms ?

full outer shell / ionic bond / electron transfer / cation / anion / charge / covalent bond / electron sharing / giant structure / molecular structure

*Now learn how to use this knowledge*

# Bonding

## Use your knowledge

20 minutes

**1** Sodium chloride contains ionic bonding.

a) Explain how ionic bonding occurs. _____ *Hint 1*
_____

b) Why are positive and negative ions formed? _____ *Hint 2*
_____

c) What holds the ions together? _____ *Hint 3*
_____

**2** Explain why potassium forms a +1 ion but oxygen forms a -2 ion. *Hint 4*
_____
_____
_____

**3** Give the charge and electronic structure of the ions listed below. *Hint 5*
lithium, Li _____
sulphur, S _____

**4** Fluorine, F, forms ionic bonds with A, a **group 2 metal.**

a) What would the charge be on the A cation? *Hint 6*
_____

b) What would the formula of the compound be? *Hint 7*
_____

**5** Chlorine reacts with carbon to form a covalent substance.

a) In terms of **electronic structure** explain how many covalent bonds carbon atoms usually form. *(Hint 8)*

_____

_____

_____

b) Suggest a formula for the substance. _____ *(Hint 9)*

**6** A student carried out several experiments to find out what types of structure and bonding are present in three substances: X, Y, and Z. The results are shown in the table below.

|  | X | Y | Z |
|---|---|---|---|
| **Solubility in water** | Insoluble | Slightly soluble | Very soluble |
| **Electrical conductivity when melted** | No | No | Yes |
| **Melting point** | Very high | Low | High |

a) Which substance has ionic bonding? _____ *(Hint 10)*

b) Explain your answer in terms of the results. *(Hint 11)*

_____

_____

_____

c) Which substance has a molecular covalent structure? *(Hint 12)*

_____

d) Explain your answer in terms of the results. *(Hint 13)*

_____

_____

_____

*Hints and answers follow*

# Bonding

1. What happens to the electrons in the sodium and chlorine atoms during bonding?

2. Atoms are gaining and losing negatively charged electrons.

3. Learn this.

4. How many electrons do the atoms have in their outer shell and how many do they want?

5. How many electrons does each atom gain or lose?

6. Consider which group A is in.

7. How many fluorine ions are needed to balance the charge on the A ion?

8. How many extra electrons does carbon need to gain?

9. How many chlorine atoms does one carbon atom need to bond with?

10. What physical properties do ionic substances have?

11. What is the structure of an ionic substance?

12. How do molecular structures differ from giant structures?

13. What kind of forces exist between separate molecules?

## Answers

1 a) electron transferred from Na to Cl b) loss of electrons gives +ve charge, gain of electrons gives -ve charge c) oppositely charged ions attract each other 2 both atoms want 8 electrons in outer shell. K has only 1 electron in outer shell which it loses to form +1 ion, O has 6 electrons in outer shell so it gains 2 electrons to form -2 ion 3 $Li^+ = 2$, $S^{2-} = 2,8,8$ 4 a) $A^{2+}$ b) $AF_2$ 5 a) both atoms need 8 electrons in outer shell, carbon has 4 electrons in outer shell therefore forms 4 bonds, chlorine has 7 electrons so forms 1 bond b) $CCl_4$ 6 a) Z b) conducts electricity in solution and when melted because of free ions, charged ions allow dissolution in water c) Y d) insoluble in $H_2O$ and doesn't conduct electricity as no ions, low melting point due to weak forces between molecules

# The periodic table

**1** The elements in the periodic table are arranged in horizontal rows called _____ and vertical columns called _____ .

**2** Most elements are metals. They usually have _____ melting and boiling points, are good conductors of _____ and _____ and are physically _____ . The other elements are called _____ . They **mostly** have _____ melting and boiling points and are **usually** physically _____ .

**3** Elements are placed in order of increasing _____ . Going across a period the _____ is filled up with electrons.

**4** Elements in the same group have similar _____ properties because they have the same _____ structure.

**5** Going down the group there is a _____ change in physical and chemical properties.

*Answers*

1 periods / groups  2 high / heat / electricity / strong / non-metals / low / weak  3 atomic number / outer shell  4 chemical / electronic  5 gradual

# The periodic table

## Improve your knowledge

**1** The periodic table drawn below shows some of the common elements.

| 1 | 2 | | | | | 3 | 4 | 5 | 6 | 7 | 0 |
|---|---|---|---|---|---|---|---|---|---|---|---|
| H | | | key | | | | | | | | He |
| Li | Be | | ☐ metal | | | B | C | N | O | F | Ne |
| Na | Mg | | | | | Al | Si | P | S | Cl | Ar |
| K | Ca | | ☐ non-metal | | | | | | | Br | Kr |
| Rb | | | Transition metals | | | | | | | I | Xe |
| Cs | | | | | | | | | | | Ra |

The elements are arranged in horizontal rows called **periods** and vertical columns called **groups**, numbered from 1 to 0 as shown (the 8th group is called group 0).

**2** More than ³/₄ of the elements are **metals**, the rest are called **non-metals** and are found in the top right hand corner (and hydrogen, H). The metals in the middle of the periodic table are called **transition metals**. Properties of metals and non-metals are shown below.

| **Metals** | **Non-metals** |
|---|---|
| Solid at room temperature (except mercury) | Many are gases |
| Most have high melting and boiling points | Most have low melting and boiling points |
| Shiny when freshly cut or scratched | Usually dull |
| Good conductors of electricity and heat | Poor conductors of heat and electricity |

| Metals | Non-metals |
| --- | --- |
| Mostly strong, but easily bent into shapes | Mostly weak and brittle |
| Form ionic bonds with non-metals | Form ionic bonds with metals and covalent bonds with other non-metals |
| Oxides usually dissolve in water to form alkaline solutions | Oxides usually dissolve in water to form acid solutions |

**3** The elements are placed in order of **increasing atomic number** from left to right. As we go across a period, an extra proton is added to the nucleus and an extra electron in the outer shell. Going from one end of a period to the other corresponds to filling the outer shell with electrons, e.g. from sodium, Na **2,8,1** to argon, Ar **2,8,8**.

**4** Elements in the same group have the same outer shell electronic structure and so elements in the same group have similar chemical properties. Group 1 elements are all metals (except hydrogen) which form +1 ions, and group 7 elements are all non-metals which form -1 ions.

**5** There is a **gradual** trend in the properties of the elements as we go down the group. Going down group 1, melting points decrease but reactivity increases. Going down group 7, melting points increase but reactivity decreases.

## Check list

**Are you sure you understand all these key ideas?**

**periodic table / group / period / metal / non-metal / outer shell electronic structure / trends down a group**

*Now learn how to use this knowledge*

# The periodic table
## Use your knowledge

20 minutes

**1** Francium, Fr, comes at the bottom of group 1. Chemists can predict some of its properties by looking at the properties of the other elements in group 1 and the trends down the group.

a) Is francium a metal or non-metal?_____ (Hint 1)

b) What kind of bonding is present in francium chloride?
_____ (Hint 2)

c) Write the formula for this compound.
_____ (Hint 3)

d) Potassium and lithium are elements in group 1. Potassium melts at 64°C and lithium melts at 180°C. Would francium melt at a higher or lower temperature than potassium? (Hint 4)
_____

e) Lithium and potassium react with cold water. Lithium reacts quite slowly but potassium reacts very quickly and gives out a lot of heat. Do you think francium would be more or less reactive than potassium? _____ (Hint 5)

f) When francium oxide dissolves in water would the solution be acidic or alkaline?
_____ (Hint 6)

**2** A chemist investigates two unknown solid elements. One is a metal and one is a non-metal.

a) The chemist cuts the elements. How might he tell them apart by their appearance?_____
_____ (Hint 7)

22

b) The chemist then burns the elements in air. Both elements form oxides which dissolve in water to form solutions. How could the chemist test the solutions to show which one was from the non-metal oxide and what would you see?   *Hint 8*

_____

_____

_____

c) Finally, the elements are reacted with chlorine and both form chlorides. One chloride is a liquid and one is a solid. The solid dissolves in water to form a solution. Which element formed the solid, the metal or non-metal? Explain your answer in terms of bonding.   *Hint 9*

_____

_____

_____

d) What other simple **physical property** could be measured to help identify the metal element?_____   *Hint 10*

**3** The element astatine, At, comes at the bottom of group 7.

a) How many electrons does astatine have in its outer shell?   *Hint 11*

_____

b) What would be the physical state of astatine at room temperature?_____   *Hint 12*

c) Would you expect astatine to be more or less reactive than chlorine?_____   *Hint 13*

d) What would the formula of the compound hydrogen astatide be? (Hydrogen astatide contains only hydrogen and astatine.)   *Hint 14*

_____

*Hints and answers follow*

# The periodic table

1. What are the other elements in the group?

2. Francium has reacted with a non-metal.

3. What groups are the two elements in and what are their valencies?

4. What is the trend down the group?

5. Is potassium more or less reactive than lithium? What is the trend?

6. What kind of solution does sodium oxide form?

7. What do metals look like when freshly cut?

8. Metal and non-metal oxides form solutions with different pHs.

9. What kind of bonding do metals and chlorine form? What are these substances like?

10. What other property, besides physical strength, do metals have which non-metals do not have?

11. What group is astatine in?

12. What is the trend from fluorine to iodine?

13. Is iodine more or less reactive than chlorine?

14. What groups are the two elements in?

*Answers*

1 a) metal b) ionic c) FrCl d) lower (<64°C) e) more reactive f) alkaline. 2 a) metal appears shiny, non-metal appears dull b) non-metal oxide would turn blue litmus paper red as it is acidic c) metal chloride is the solid because metal chloride contains ionic bonding and ionic substances have high melting points (or are solids at room temperature) d) test to see if elements conduct electricity - metals do, non-metals usually do not 3 a) 7 b) solid c) less reactive d) HAt

# Metals and the reactivity series

## Test your knowledge

**1** Compare the following properties of group 1 metals with those of iron. Melting points are _____ . Reactivity is _____ .

**2** Sodium reacts with water to form _____ and _____ gas.

**3** What type of bonding is present in potassium oxide? _____

**4** Is a solution of potassium oxide acidic or alkaline? _____

**5** Transition metals have _____ melting points than group 1 metals. They are good conductors of _____ and _____ .

**6** We can often tell that a compound contains a transition metal from its appearance, because the compounds are often _____ .

**7** Which is more reactive – calcium, Ca, or iron, Fe? _____

**8** Zinc is below magnesium in the reactivity series. Would zinc displace magnesium from a solution of magnesium chloride? _____

## Answers

**1** lower / greater **2** sodium hydroxide / hydrogen **3** ionic **4** alkaline **5** higher / electricity / heat **6** coloured **7** calcium **8** no

# Metals and the reactivity series

## Improve your knowledge

20 minutes

**1** Metals in group 1 are **soft, light**, have **low melting and boiling points** and are **very reactive**, e.g. lithium, Li, sodium, Na and potassium, K. Going **down** the group, the metals become lighter, softer, have lower melting and boiling points and become **more** reactive.

**2** Group 1 metals react with water to give a solution of the **hydroxide** and **hydrogen gas**, e.g. Na(s) + $H_2O$(l) → NaOH(aq) + $H_2$(g)

Hydrogen gas '**pops**' with a lighted splint - this is a **test**.

**3** Group 1 metals react with non-metals to form **ionic solids** containing the metals as **+1 ions**. e.g. 2Li(s) + $Cl_2$(g) → 2LiCl(s)
4Na(s) + $O_2$(g) → 2$Na_2$O(s)

**4** Group 1 metal **oxides** and **hydroxides** dissolve to give **alkaline** solutions which turn red litmus paper **blue**.

**5** Transition metals are **hard, dense**, have **high melting and boiling points** and are **good conductors** of **electricity** and **heat**. They are **less reactive** than group 1 metals, reacting slowly with water.

| Copper, Cu | Good conductor of electricity - used for electrical wires |
|---|---|
| Iron, Fe | Strong - used for construction, e.g. cars, buildings, ships |
| Nickel, Ni | Used to make coins |

Note that iron is usually mixed with other metals to form the alloy steel.

**6** Many transition metal compounds are **coloured**, e.g. copper sulphate is blue and iron oxide is brown.

**7** Some metals are more reactive than others. Metals can be placed in order of reactivity. This is the **reactivity series** and must be learnt.

| Metal | Symbol | Notes |
|---|---|---|
| Potassium | K | All react vigorously with water to form hydroxides and hydrogen, and with air to form oxides |
| Sodium | Na | |
| Calcium | Ca | |
| Magnesium | Mg | Reacts with steam, but very slowly with cold water |
| Aluminium | Al | React with steam only when heated |
| Zinc | Zn | |
| Iron | Fe | Reacts very slowly with steam when heated |
| (Hydrogen) | ($H_2$) | |
| Copper | Cu | Elements less reactive than $H_2$ don't react with acids |

**8** A **more** reactive metal will **displace** a **less** reactive metal from its compounds in solution, e.g. zinc metal will displace copper from copper sulphate solution.   $Zn(s) + CuSO_4(aq) \rightarrow ZnSO_4(aq) + Cu(s)$

# Check list

**Are you sure you understand the following key terms?**

**group 1 metal / reactions of group 1 metals with water and other non-metals / alkaline oxide / transition metal / reactivity series / displacement reaction**

*Now learn how to use this knowledge*

# Metals and the reactivity series

## Use your knowledge

**1** Rubidium, Rb, is a group 1 metal. It is stored under oil to keep it fresh. A chemist found an old jar of rubidium from which most of the oil had leaked and the metal was covered in a white powdery substance.

a) Why is rubidium stored under oil?  *(Hint 1)*

b) What is the white powder and what is its formula?  *(Hint 2)*

c) Why couldn't you store rubidium under water?  *(Hint 3)*

d) Write a word equation showing what would happen if rubidium was stored in water.  *(Hint 4)*

e) Rubidium chloride is also a white powder and looks like the substance described above. By dissolving the two substances in water, separately, describe how the chemist could tell the two substances apart.  *(Hint 5)*

**2** Old iron garden gates are often covered in a flaky, brown substance and have holes where this has formed.

a) What is the brown substance? How does it form?  *(Hint 6)*

**2** b) Why is this a problem? _____  *Hint 7*
_____

c) Write a word equation for this reaction.  *Hint 8*
_____

d) Stainless steel is resistant to rusting. Why is it used to make kitchen utensils?  *Hint 9*
_____

e) Why are transition metals used in the manufacture of paints?  *Hint 10*
_____

**3** A student carried out experiments to find out where the metal tin comes in the reactivity series. After each observation below state where it places tin in the series.

a) Tin did not react with cold water.  *Hint 11*
_____

b) Tin reacted **slowly** with steam **only** when strongly heated.  *Hint 12*
_____

c) Tin reacted with hydrochloric acid to form hydrogen.  *Hint 13*
_____

d) There was **no** reaction between tin and iron sulphate solution.  *Hint 14*
_____

*Hints and answers follow*

# Metals and the reactivity series

## Hints

1. Rubidium is very reactive. What might it easily react with?

2. What is the valency of the element that rubidium may have reacted with?

3. What do you know about group 1 metals and water?

4. One of the products is hydrogen.

5. Rubidium chloride forms a neutral solution.

6. What is in the air which iron reacts with?

7. The brown substance is not very strong.

8. What are the reactants? Only one product is formed.

9. Kitchen utensils are used to prepare food and for eating.

10. How is the appearance of transition metals different from most compounds?

11. How far down the series do metals react with cold water?

12. It must be less reactive than metals that will react easily with steam, when heated.

13. Metals that dissolve in acids are displacing hydrogen.

14. Which metal must be most reactive?

## Answers

**1** a) to prevent the rubidium reacting with the oxygen in the air b) rubidium oxide, $Rb_2O$ c) rubidium would react with water d) rubidium + water → rubidium hydroxide e) solution of rubidium oxide would be alkaline and turn red litmus paper blue, rubidium chloride would be neutral and show no colour change with litmus paper **2** a) rust or iron oxide, iron reacts with oxygen in the air b) rust is weak and makes the gate more likely to break etc c) iron + oxygen → iron oxide d) do not want rust to get into food and be eaten e) coloured compounds **3** a) below magnesium b) below zinc c) above $H_2$ d) below iron

# Non-metals

## Test your knowledge

**1** Describe the appearance of bromine at room temperature.
_____  _____

**2** Halogens react with _____ to form compounds with ionic bonds. The halogen is present as an ion with a charge of _____ .

**3** Hydrogen chloride contains _____ bonds. It dissolves in water to form a solution with an _____ pH.

**4** Fluorine will _____ bromine from solutions of its compounds.

**5** Chlorine turns damp, blue litmus paper _____ then _____ it.

**6** Chlorine is added to drinking water because it kills _____ .
Fluorine is added to drinking water because it strengthens _____ .

**7** Group 0 elements are all in the _____ state at room temperature.

**8** Group 0 elements are inert because they have _____ in their outer shells.

**9** Neon is used in luminous _____ .

## Answers

1 red-brown liquid  2 metals / –1  3 covalent / acidic  4 displace  5 red / bleaches  6 bacteria / teeth  7 gas  8 8 electrons  9 signs or lamps

31

# Non-metals

## Improve your knowledge

**1** Elements in group 7 are called the **halogens.** They are **molecular covalent** non-metals, with each molecule containing two atoms.

| Fluorine | Chlorine | Bromine | Iodine |
|---|---|---|---|
| $F_2$ | $Cl_2$ | $Br_2$ | $I_2$ |
| Pale yellow gas | Pale green gas | Red / brown liquid | Grey - black solid |
| Melting and boiling points increase down the group ||||
| Most reactive | Reactivity decreases down group || Least reactive |

**2** Halogens react with metals to form **ionic** salts containing the halogen as a **-1 halide** ion.
e.g.  2K(s)    +  $Br_2$(g)   →   2KBr(s)
      potassium + bromine  →  potassium bromide

**3** Halogens react with non-metals to form molecular covalent compounds.
e.g.  $F_2$(g)   +  $H_2$(g)   →   2HF(g)
      fluorine  +  hydrogen  →   hydrogen fluoride
The hydrogen halides (including hydrogen chloride, bromide and iodide) are soluble in water and give **acidic** solutions which turn blue litmus paper red. The hydrogen halide **gases** turn damp, blue litmus paper red.

**4** A **more** reactive halogen will **displace** a **less** reactive halogen from its compounds. For example, bromine will displace iodine from a solution of sodium iodide.
$$Br_2(l) + 2NaI(aq) → 2NaBr(aq) + I_2(s)$$

**5** Chlorine can be identified by using damp, blue litmus paper. The paper will first turn red and then white as it is **bleached.**

**6** Some uses of halogens and their compounds are described below.

| Element / Compound | Use | Reason |
|---|---|---|
| Fluorine | Toothpaste/drinking water | Strengthens teeth |
| Chlorine | Bleach<br>Drinking water | Bleaching agent<br>Kills bacteria |
| Iodine | Antiseptic | Kills bacteria |
| Silver halides e.g. AgI | Photography | Change colour when exposed to light |

**7** Elements in group 0 are called the **noble gases**. They include: Helium, He; Neon, Ne; Argon, Ar; Krypton, Kr; Xenon, Xe; Radon, Ra. They are all gases with very low boiling and melting points.

**8** Because noble gases have 8 electrons in the outer shell, they are **unreactive** or **inert**. They are made up of single, un-bonded atoms, not molecules. They are described as **monatomic**.

**9** Noble gases are used in street lamps and discharge tubes. A tube filled with gas has an electrical current passed through it which causes a bright, luminous glow, such as in neon signs.

## Check list

**Are you sure you understand the following key terms?**

**halogen / halogen reactions with metals / halogen reactions with non-metals / displacement reactions / tests for chlorine / uses of halogens / noble gas / lack of reactivity of noble gases / uses of noble gases**

*Now learn how to use this knowledge*

# Non-metals

**1** a) What kind of bonding does phosphorus chloride contain?

*Hint 1*

Phosphorus chloride reacts with water giving off fumes of hydrogen chloride. This gas dissolves in water forming a solution. This solution reacts with potassium hydroxide solution, an alkali, to give potassium chloride and water.

b) Describe a test for hydrogen chloride gas and what you would see.

*Hint 2*

c) Why does the solution react with potassium hydroxide?

*Hint 3*

d) Write a word equation for this reaction, using state symbols.

*Hint 4*

**2** Bromine reacts with calcium, Ca, to form a white solid R. R dissolves in water to produce a colourless solution. When chlorine gas is bubbled through the solution a brown liquid, X, was produced and a **new** colourless solution, Y.

a) What type of bonding is present in R?

*Hint 5*

b) Identify R and give its formula.

*Hint 6*

c) Explain what happens when the chlorine gas is bubbled through the solution of R, and therefore identify X and Y.

*Hint 7*

**3** Many years ago the water supply was not chemically treated and disease and tooth decay were common. Explain why the addition of chlorine and fluorides has reduced the occurrence of these medical problems. *(Hint 8)*

_____

_____

_____

_____

**4** The noble gases are un-reactive, monatomic gases.

a) Explain what is meant by a monatomic gas. *(Hint 9)*

_____

b) How many electrons do noble gases have in their outer shells? *(Hint 10)*

_____

c) By thinking about their electronic structures, explain why the noble gases are so un-reactive. *(Hint 11)*

_____

_____

_____

**5** Hydrogen gas is very light and was used to fill 'lighter-than-air balloons' and airships. However it was not very safe. These days helium is often used, even though it is not as light as hydrogen.

a) What do you think was the danger of using hydrogen? *(Hint 12)*

_____

_____

b) Why is helium used instead even though it is heavier? *(Hint 13)*

_____

_____

*Hints and answers follow*

# Non-metals

## Hints

1. Phosphorus is a non-metal.
2. Learn this.
3. What type of solutions do hydrogen halides form in water?
4. What are the reactants and products?
5. Ca is a metal.
6. What groups are the elements in - what are their valencies?
7. Chlorine is more reactive than bromine.
8. What causes the diseases and what do the halogens do?
9. 'Mono' means one.
10. Look at the position of the group in the periodic table.
11. How many electrons do atoms need in their outer shells?
12. Hydrogen burns in air releasing a lot of energy.
13. What would happen if helium and air mixed?

## Answers

1 a) covalent b) place damp, blue litmus paper in the gas, it turns red c) because hydrogen chloride solution is acidic and neutralises the alkali d) hydrogen chloride (aq) + potassium hydroxide (aq) → potassium chloride (aq) + water (l) 2 a) ionic b) $CaBr_2$ c) chlorine is more reactive than bromine and displaces it from its solution to form calcium chloride solution, Y, and bromine, the brown liquid X 3 chlorine kills bacteria in the water which cause diseases, fluorides strengthen teeth and help them resist tooth decay 4 a) a gas where the element exists as a single, unbonded atom b) 8 c) they already have a stable, full outer shell of electrons so they do not need to form any bonds with other atoms 5 a) it would be likely to explode or catch fire causing harm to people and damage to buildings etc b) because it is un-reactive it will not explode or react with air so it is much safer

# Rates of reactions

**1** Chemical reactions occur when particles _____ with one another, with enough _____ to form new substances.

**2** The rate of all chemical reactions can be increased by increasing the _____ or _____ . For solids the rate of reaction can also be increased by increasing the _____ _____ .

**3** Calcium carbonate reacts with acid to form carbon dioxide gas. The rate of reaction could be found by measuring the _____ of _____ at intervals over a period of time.

**4** The rate of reaction between calcium carbonate and acid can be compared for two different experiments by plotting a graph of _____ on the y-axis and _____ on the x-axis.

**5** If the same amount of calcium carbonate is reacted with dilute and concentrated acid, would the reaction with concentrated acid produce more, less or the same amount of carbon dioxide? _____

**6** A catalyst is a substance which _____ the rate of reaction, but which is not _____ _____ during the reaction.

**7** Catalysts in living cells are called _____ .

*Answers*

1 collide / energy  2 temperature / concentration / surface area  3 volume OR mass / gas  4 volume OR mass of gas / time  5 same  6 increases / used up  7 enzymes

# Rates of reactions
## Improve your knowledge

**20 minutes**

**1** Chemical reactions occur when particles **collide** with **enough energy** to form new substances. The more often successful collisions occur the faster the **rate of the reaction.**

**2** The rate of a reaction can be increased by **increasing** several factors:

| Temperature | Particles move more quickly **and** collide with more energy |
| --- | --- |
| Concentration | More particles present so they are more likely to collide |
| Surface area of solids | Easier for other particles to come into contact with the solid |

**3** The rate of reaction can be found by measuring how much of one of the **products is formed** or one of **the reactants disappears** at intervals over a period of **time**. See below.

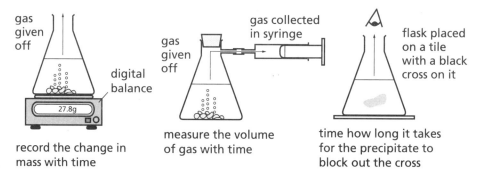

**4** The **results** of experiments can be **plotted on a graph**. The graph on page 39 shows the amount of hydrogen gas produced over a period of time, when magnesium was reacted with hydrochloric acid. Experiment 2 was carried out at a **higher temperature**.

**5** We can use the graph to find out information about the reactions.

In experiment 1, 40 cm³ of gas were produced in 24 seconds.
In experiment 2, after 70 seconds, 98 cm³ of gas were produced.

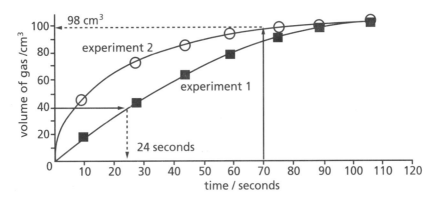

The line for experiment 2 rises **more steeply** than the line for experiment 1 at the **start,** and this means that the reaction in experiment 2 has a **faster rate.**

The **same** total volume of hydrogen gas was produced in **both** experiments because the **same** amount of magnesium and acid reacted together.

**6** A **catalyst** is a substance which **speeds up a chemical reaction** but is **not used up** in the reaction itself. Catalysts are often used in industry to speed up reactions without using **expensive** high temperatures or concentrations.

**7** Living cells use catalysts called **enzymes**. Some examples are shown below.

| Source of enzyme | What the enzyme does |
|---|---|
| Yeast | Converts sugar into alcohol. This is called fermentation |
| Yeast | Enzymes produce $CO_2$ gas which causes bread to rise |
| Bacteria | Enzymes in bacteria cells convert milk to yoghurt |

Enzymes only work at the **right** temperature and pH. If the conditions change the enzyme stops working and the reaction slows down and stops.

# Check list

Are you sure you understand the following key terms?

**collisions / rate of reaction / factors affecting rate of reaction / rate of reaction experiments / rate of reaction graphs / catalyst / enzyme / fermentation**

*Now learn how to use this knowledge*

# Rates of reactions

## Use your knowledge

**1** The graph shows the results of two experiments to measure the rates of reaction between zinc and hydrochloric acid. Both experiments were carried out using the same conditions, **except** the concentration of acid was different in the two experiments.

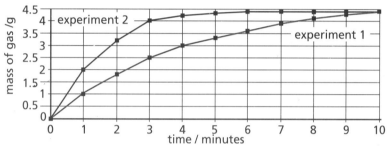

a) From the graph, find out how much hydrogen gas was produced in each experiment after 3½ minutes.

Experiment 1 _____ Experiment 2 _____

*Hint 1*

b) From the graph, find out how long it took in each experiment for 4 grams of hydrogen to be produced.

Experiment 1 _____ Experiment 2 _____

*Hint 2*

c) From your answers in a) and b) which reaction occurred at the faster rate?

_____

*Hint 3*

d) Which experiment used the more concentrated acid? Explain your answer in terms of how reactions occur.

_____

_____

_____

*Hint 4*

**2** The graph shows how the **rate** of an enzyme-catalysed reaction changes as temperature varies.

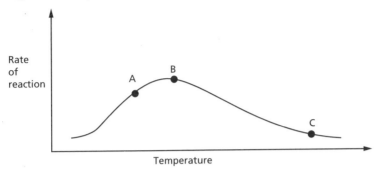

a) Between points A and B temperature increases. What happens to the rate of reaction? *(Hint 5)*

b) Explain this change in terms of the particles involved in the reaction. *(Hint 6)*

c) Between points B and C temperature increases. What happens to the rate of reaction? *(Hint 7)*

d) Why does this change occur as the temperature increases? *(Hint 8)*

**3** A chemical company, Alco, sets up a factory to produce a new plastic. Alco must produce **enough** plastic each week to be profitable. Alco's process uses an expensive catalyst.

a) Why do you think Alco uses a catalyst to be profitable? *(Hint 9)*

b) Why doesn't an expensive catalyst lead to high costs every week? *(Hint 10)*

*Hints and answers follow*

# Rates of reactions

1. Draw a line from the horizontal axis showing 'time', at 3½ minutes to the two curves.

2. Draw a line from the axis showing 'amount of gas produced' at 4 g to the two curves.

3. In which experiment was hydrogen gas produced most quickly?

4. Why does concentration affect rate of reaction?

5. Read this from the graph.

6. As temperature increases what happens to the speed of the particles?

7. Read this from the graph.

8. What happens to enzymes at high temperatures?

9. The plastic must be produced quickly enough. What does a catalyst do?

10. What happens to a catalyst during a chemical reaction?

## Answers

1 a) experiment 1: 2.7 - 2.8 g / experiment 2: 4.1 - 4.2 g b) experiment 1: 7.3 - 7.4 mins / experiment 2: 3.0 mins c) experiment 2 d) experiment 2, more particles present therefore more collisions occur leading to a higher rate of reaction. 2 a) increases b) particles move more quickly and with more energy and so there are more successful collisions which speeds up the reaction c) decreases d) enzymes are destroyed by high temperature so the reaction is not catalysed and slows down 3 a) a catalyst speeds up the reaction, producing more plastic per week b) catalysts are not used up in a chemical reaction and so only need to be bought once

# Gases and ions

## Test your knowledge

**1** When collecting gases _____ delivery is used to collect gases that are more dense than air, e.g. _____ _____ . When the volume of gas collected is required, a _____ _____ is used. Gases that are insoluble in _____ are collected _____ _____ .

**2** Hydrogen gas gives a _____ _____ when exposed to a lighted splint. Oxygen gas _____ a glowing splint. Ammonia is an alkaline gas and so turns moist pH paper _____ . The gas that turns limewater milky is _____ _____ .

**3** The presence of $H^+$ in solution can be detected by addition of _____ paper which turns red. Flame tests are used to detect the presence of some metals in compounds, e.g. if sodium is present then the flame turns _____ . To distinguish between $Fe^{2+}$ and $Fe^{3+}$ ions dilute _____ _____ solution is added. $Fe^{2+}$ forms a _____ coloured solid whereas $Fe^{3+}$ forms a _____ coloured solid.

**4** Upon addition of dilute hydrochloric acid and barium chloride solution to a solution containing sulphate ions ($SO_4^{2-}$) a _____ solid is formed. To distinguish between the halide ions ($Cl^-$, $Br^-$ and $I^-$) dilute nitric acid and _____ _____ are added. Each halide forms a different coloured solid, for example, the chloride ion forms a _____ precipitate whereas the _____ ion forms a yellow precipitate.

## Answers

**1** downward / carbon dioxide / gas syringe / water / over water **2** squeaky 'pop' / relights / blue or purple / carbon dioxide **3** pH / yellow / sodium hydroxide / green / red/brown **4** white / silver nitrate / white / iodide

If you got them all right, skip to page 46

# Gases and ions

## Improve your knowledge

**20 minutes**

**1**
a) **Upward delivery**
This method is used to collect gases **less** dense than air, e.g. ammonia ($NH_3$).

b) **Downward delivery**
This method is used to collect gases **more** dense than air, e.g. carbon dioxide ($CO_2$).

c) **Collection over water**
This method is used to collect any gas that is insoluble in water, e.g. hydrogen ($H_2$) and oxygen ($O_2$).

d) **Collection in a gas syringe**
This method is used for any gas whenever the **volume** collected is to be measured.

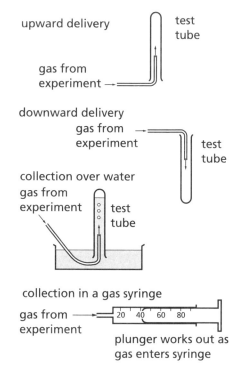

**2** As well as knowing the method of collection of the above gases, you also need to know how to identify them.

| Gas | Test | Positive result |
| --- | --- | --- |
| Hydrogen | Place **lighted** splint in gas* | Squeaky 'pop' |
| Oxygen | Place **glowing** splint in gas | Splint relights |
| Ammonia | Place damp pH paper in gas | pH paper turns blue or purple |
| Carbon dioxide | Pass gas through limewater | Limewater turns milky |

\* Mixtures of hydrogen and air form explosive mixtures.

## 3

| Cation | Test | Positive result |
|---|---|---|
| H+ | pH paper<br>Addition of a reactive metal, base or metal carbonate | Turns red<br>Salt is formed ($H_2$ evolved) $CO_2$ given off |
| Na+ | Flame test | Yellow flame |
| K+ | Flame test | Lilac flame |
| $Ca^{2+}$ | Flame test | Red flame |
| $Ca^{2+}$ | Addition of sodium hydroxide solution | White solid (if conc NaOH) |
| $Cu^{2+}$ | Addition of sodium hydroxide solution | Blue solid: $Cu(OH)_2$ |
| $Fe^{2+}$ | Addition of sodium hydroxide solution | Green solid: $Fe(OH)_2$ |
| $Fe^{3+}$ | Addition of sodium hydroxide solution | Red/brown solid: $Fe(OH)_3$ |
| $NH_4^+$ | Addition of sodium hydroxide solution and warm | Ammonia given off |

## 4

| Anion | Test | Positive result |
|---|---|---|
| OH− | Add pH paper<br>Heat with an ammonium salt | Turns green/blue<br>Ammonia is given off |
| $SO_4^{2-}$ | Add barium chloride solution and dilute hydrochloric acid | White solid formed $BaSO_4$ |
| $CO_3^{2-}$ | Add dilute acid and bubble gas collected through limewater | Limewater turns milky |
| Cl− | Add silver nitrate solution and dilute nitric acid | White precipitate |
| Br− | Add silver nitrate solution and dilute nitric acid | Creamy precipitate |
| I− | Add silver nitrate solution and dilute nitric acid | Yellow precipitate |

## Check list

Are you sure you understand the following key terms?

methods of collecting gases / how to collect and identify $NH_3$, $CO_2$, $H_2$ and $O_2$ / tests for cations / tests for anions

*Now learn how to use this knowledge*

# Gases and ions
## Use your knowledge

20 minutes

**1** When an unknown solid A was analysed using a flame test a yellow flame was observed. In a reaction with dilute acid a gas was given off that turned limewater milky.

   a) Suggest the identity of the metal in A. _____     *Hint 1*

   b) What evidence tells you this? _____

   c) What gas does A give off upon reaction with acid?
   _____     *Hint 2*

   d) How could you collect this gas? _____     *Hint 3*

   e) Suggest the identity of the anion in A. _____

   f) Suggest the formula of A. _____     *Hint 4*

**2** An unknown solution B turns pH paper red, and, upon reaction with zinc, 37cm³ of gas was collected. To B, dilute hydrochloric acid and another liquid labelled C was added and immediately a white solid was observed.

   a) What type of substance turns pH paper red? _____     *Hint 5*

   b) What ion does this tell you is present in B? _____     *Hint 6*

   c) What would you use to collect the gas given off?
   _____     *Hint 7*

   d) How could you prove whether the gas given off was hydrogen?     *Hint 8*
   _____
                                                                 *Hint 9*
   e) Suggest the identity of solution C. _____

   f) What ion does this test for? _____     *Hint 10*

   g) Suggest the identity of B. _____

**3** A student was asked to analyse solid D and solution E. The instructions on the practical sheet asked the student to carry out a flame test on D and to add dilute sodium hydroxide to E.

a) Solid D gave a red flame in the flame test. What metal does this suggest is present?  *(Hint 11)*

_____

b) How else could the student test for this ion?  *(Hint 12)*

_____

c) Upon addition of dilute sodium hydroxide to E, a red/brown solid was formed. What cation does this suggest is present in E?  *(Hint 13)*

_____

**4** Another student was analysing the anions present in D and E. He dissolved D in water to which he added pH paper, which immediately turned dark blue.

a) Suggest the formula of the anion in D.  *(Hint 14)*

_____

b) Suggest the formula of D.

_____

To E the student added silver nitrate solution and dilute nitric acid. E contains chloride ions.

c) What will the student have observed upon addition of the silver nitrate and nitric acid?  *(Hint 15)*

_____

d) What would the student have observed if bromide ions had been present instead of chloride ions?  *(Hint 16)*

_____

e) What is the formula of E?

_____

*Hints and answers follow*

# Gases and ions

## Hints

1. The metal is identified using a flame test.

2. Which gas turns limewater milky?

3. The gas given off is denser than air.

4. Which anion gives off this gas when acid is added?

5. It is either an acid or an alkali.

6. Which ion is responsible for the answer to part 2 a)?

7. The exact volume of gas collected is known.

8. What is the chemical test for hydrogen gas?

9. Which test involves adding dilute hydrochloric acid, then another solution, and results in the observation of a white precipitate?

10. Which anion does this test for?

11. Learn the results of flame tests!

12. What is the additional test for this anion?

13. This is a test for an 'iron' cation.

14. Which ion is this the chemical test for?

15. What is observed when these solutions are added to chloride ions?

16. How can you use this test to distinguish between chloride and bromide ions?

## Answers

**1** a) sodium b) yellow flame c) carbon dioxide d) downward delivery e) $CO_3^{2-}$ f) $Na_2CO_3$ **2** a) acid b) $H^+$ c) gas syringe d) there would be a squeaky 'pop', upon exposure to a lighted splint e) barium chloride solution f) sulphate anion ($SO_4^{2-}$) g) sulphuric acid ($H_2SO_4$) **3** a) calcium b) add concentrated sodium hydroxide solution and observe a white solid c) $Fe^{3+}$ **4** a) hydroxide ($OH^-$) ion b) $Ca(OH)_2$ c) a white precipitate d) a creamy precipitate e) $FeCl_3$

# Water

## Test your knowledge

**1** When soap is used in soft water a _____ is formed, whereas in hard water a _____ is formed. Hard water is caused by the presence of dissolved _____ and _____ salts. These salts are dissolved in rainwater that passes over gypsum, limestone and _____ . There are two types of hard water, _____ and _____ hardness.

**2** _____ hardness can be removed by boiling the water. However _____ hardness can only be removed by adding _____ _____ or by passing the water through an _____ _____ column. Hard water can cause problems to the inner surface of pipework because it can deposit _____ which can cause _____ . However, in copper and lead water pipes the layer of _____ can _____ the entry of these toxic metals into the water.

**3** Before water is suitable for human consumption it needs to be treated. The first step is to allow any large solids in water to settle out. This process is called _____ . The water is then _____ through sand to remove any small particles before being _____ to kill any bacteria present. Often _____ is added to water to help produce stronger teeth.

## Answers

*(printed upside down)*

**1** lather / scum / calcium and magnesium / chalk / temporary and permanent
**2** temporary / permanent / sodium carbonate / ion exchange / scale / blockages / scale / prevent/reduce
**3** sedimentation / filtered / chlorinated / fluoride

If you got them all right, skip to page 52

49

# Water

## Improve your knowledge

**1** Formation of hard water

Rain often falls on limestone, chalk and gypsum rocks. This water then dissolves calcium and magnesium salts and is called **hard water.** Hard water is defined as water that doesn't form lather with soap, but forms a scum instead. The dissolved compounds in hard water react with the soap to form scum.

Thus using soap in hard water areas is wasteful (because of the lack of lather formed). However, using soapless detergents (e.g. shampoo) is much less wasteful because they are not affected by hard water.
There are two types of hardness:

- **Temporary hardness** (caused by dissolved calcium and magnesium hydrogencarbonate salts).

- **Permanent hardness** (caused by dissolved chlorides and sulphates of magnesium and calcium).

**2** Softening water (i.e. removing hardness)

All hardness can be removed from water by distilling it. However, this is an uneconomical process and other (cheaper) methods are used instead. Temporary hardness is removed simply by boiling. There are two methods of removing permanent hardness:

- **Addition of water softening chemicals** e.g. sodium carbonate crystals (this precipitates out the magnesium and calcium as magnesium carbonate and calcium carbonate).

- **Use of an ion exchange resin** (this replaces the calcium and magnesium ions in water with sodium ions which do not cause hardness).

The table below gives the advantages and disadvantages of hard water:

| Advantages | Disadvantages |
|---|---|
| Contains calcium compounds that are required by the body for strong bones and teeth | Wastes soap because of the formation of scum, which can lead to increased costs |
| Formation of scale on the inner surface of copper and lead pipes reduces the entry of poisonous lead and copper salts into water. Scale also prevents corrosion of the pipes | Forms a layer of 'fur' in kettles and scale in boilers and pipes. This scale may lead to blocked pipes and reduced efficiency of radiators |

3. Water treatment

Water needs to be treated before it is fit for human consumption. There are three main stages in the treatment of water:

- **Sedimentation** Water (stored in a reservoir) is transferred to a large tank where it is allowed to settle. This allows any solid impurities to sink to the bottom (forming sludge).

- **Filtration** The water is now allowed to pass through sand beds that filter out small particles.

- **Chlorination** Chlorine is now added to the water to kill any bacteria present. The water is now safe to drink.

Often, calcium fluoride is added to drinking water. This process is known as **fluoridation**. This is done because the presence of fluoride ions in water produces stronger teeth which are more resistant to decay.

# Check list

**Are you sure you understand the following terms?**

**formation of permanent and temporary hardness / methods of softening water / stages in water treatment**

*Now learn how to use this knowledge*

# Water

**1** Samples of water (A, B and C) from three different areas were analysed in a laboratory to determine how hard they were. The tests and their results are shown in the table below.

| Test | A | B | C |
|---|---|---|---|
| Shaken with soap solution | Scum | Lather | Scum |
| Boiled first and then shaken with soap solution | Lather | Lather | Scum |

a) One of the samples is distilled water. Which one do you think this is? Explain your answer. *(Hint 1)*

_____

_____

The other two samples were taken from hard water areas.

b) Which sample contains temporary hardness? _____ *(Hint 2)*

c) Which sample contains permanent hardness? _____

**2** Soap solution was added to four samples of water from different areas to see whether lather or a scum was formed. This experiment was repeated using more of the samples of water after they had been boiled, and a third time using samples that had been treated with sodium carbonate crystals. The results are shown below.

| Sample | Untreated | Boiled | Treated with sodium carbonate crystals |
|---|---|---|---|
| A | Scum | Lather | Lather |
| B | Scum | Scum | Lather |
| C | Lather | Lather | Lather |

a) Which of the samples do you think is the hardest water? Explain your answer. *(Hint 4)*

_____

_____

b) Name a chemical which could be responsible for the hardness in the samples below.

Sample A _____

Sample B _____ *(Hint 5)*

c) The water treated with sodium carbonate crystals was filtered before being used in the experiment. What was removed by filtering? *(Hint 6)*

_____

d) How does this treatment reduce hardness? *(Hint 7)*

_____

**3** Vanessa was investigating the purification of water by observing pond water. She collected a sample of pond water in a jam jar and left it on her bedroom window overnight. The next morning she noticed that the water was a lot clearer and that there was a layer of sludge on the bottom of the jam jar.

a) What stage in the purification of water does this represent? *(Hint 8)*

_____

b) Vanessa noticed that there were still lots of small impurities in the water. How do water companies remove small particles from water? *(Hint 9)*

_____

c) The water now looks as clear as tap water and Vanessa is tempted to drink it. Why shouldn't she drink this water? *(Hint 10)*

_____

*Hints and answers follow*

# Water

*Hints*

1. Distillation removes hardness from water.

2. How is temporary hardness removed from water?

3. How does permanently hard water react with soap? How does boiling affect permanently hard water?

4. How would you expect hard water to react in the first two experiments?

5. You need to learn the names of the salts that cause temporary and permanent hardness.

6. What happens when sodium carbonate crystals are added to hard water?

7. What is removed from the water in the filtrate?

8. One of the stages of water purification involves leaving the water to settle.

9. This process involves passing the water through sand.

10. What else is done to the water before it is delivered to our homes?

*Answers*

1 a) sample B because it forms a lather with soap before it is boiled b) sample A (because the hardness is removed by boiling) c) sample C (because it doesn't form a scum even after boiling) 2 a) sample B because only when sodium carbonate is added does it form a lather b) sample A: calcium hydrogencarbonate or magnesium hydrogencarbonate / sample B: magnesium chloride or magnesium sulphate or calcium chloride or calcium sulphate c) calcium carbonate and/or magnesium carbonate d) The calcium and/or magnesium ions are removed from the water as an insoluble precipitate 3 a) sedimentation b) by filtering the water through sand c) the water may contain harmful bacteria (the water needs to be chlorinated before it is safe for consumption)

# Salts – preparation and solubility

## Test your knowledge

1. A _____ is a substance that is dissolved in a liquid. The liquid is called a _____ . When no more solid will dissolve in a liquid then a _____ _____ has been formed. How soluble a substance is depends not only on what the substance is but also on the _____ and _____ .

2. Complete the table below showing the solubilities of some common salts. The first one has been done for you.

| Substance | Soluble |
|---|---|
| sodium chloride | ✓ |
| barium sulphate | |
| magnesium chloride | |
| potassium nitrate | |
| silver chloride | |

3. Soluble salts can be prepared by reacting _____ with bases, metals or metal carbonates. The salt can be obtained by _____ the water from the resultant solution.

4. Insoluble salts are prepared as _____ i.e. a solid formed when two liquids react. The solid is obtained by _____ .

## Answers

1 solute / solvent / saturated solution / solvent / temperature / pressure  2 ✗ / ✓ / ✓ / ✗  3 acids / evaporating  4 precipitates / filtration

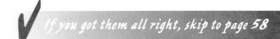

If you got them all right, skip to page 58

55

# Salts – preparation and solubility

## Improve your knowledge

**1** You need to be familiar with the following terms related to solubility.

**Solute** – This is the substance that is dissolved.

**Solvent** – This is the liquid in which the solute is dissolved.

**Solution** – This is the mixture of the dissolved solute and solvent.

**Aqueous** – A solute dissolved in water is referred to as an aqueous solution.

**Saturated solution** – This is a solution in which no more solute will dissolve.

When a solute is dissolved in a solvent, a point will be reached where no more solid dissolves, i.e. the solution is saturated. The mass of solute that will dissolve in a given volume of solvent depends upon certain factors.

- What the solvent is
- What the solute is
- The temperature of the solvent

**2** Solubility curves

The maximum mass of a solid that can be dissolved in 100 cm³ of water at different temperatures can be determined and plotted. This is called a solubility curve and is shown below for potassium nitrate.

The curve represents the maximum mass of potassium nitrate that will dissolve in 100 cm³ of water at a given temperature. This forms a saturated solution.

It should be noted that the solubility of gases in liquids decreases with increasing temperature.

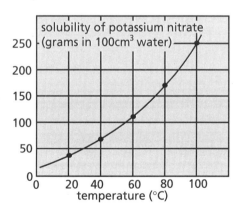

## 3 Rules of solubility

You need to learn the following rules of solubility for different salts in water.

| Salt | Soluble |
|---|---|
| All Na, K and $NH_4^+$ salts | ✓ |
| Nitrates e.g. lithium nitrate | ✓ |
| Common chlorides (except Ag and Pb) | ✓ |
| Common sulphates (except Pb, Ba and Ca) | ✓ |
| Common carbonates and hydroxides – | ✗ |
| except for those of Na, K, $NH_4^+$ | ✓ |

Note: Salts that are referred to as insoluble are, in fact, slightly soluble.

## 4 Preparation of salts

The following reactions of acids produce **soluble** salts:

- acid + base → salt + water
- acid + metal → salt + hydrogen
- acid + metal carbonate → salt + carbon dioxide + water

The salts prepared from these reactions of acids can be obtained by evaporating off the water from the product. To obtain a pure product it is necessary when carrying out the reaction to use an excess of the solid. This prevents any acid being left over. The remaining solid can be removed by filtration.

Insoluble salts can be formed as precipitates, i.e. a solid formed when two liquids react. A pure, dry sample of the salt can be obtained by filtering the precipitate, rinsing with water and then evaporating the water off (by placing the salt in a hot oven, above 100°C).

## Check list

**Are you sure you understand the following key terms?**

**solubility definitions / interpretation of solubility curves / rules of solubility / how to prepare soluble and insoluble salts**

*Now learn how to use this knowledge*

# Salts – preparation and solubility

## Use your knowledge

**1** A student dissolved copper sulphate in water at 30°C until no more would dissolve. The water was then heated to 50°C and the student noted that more copper sulphate dissolved.

In the experiment above name the solute and solvent.

a) The **solute**: _____   *Hint 1*

b) The **solvent**: _____   *Hint 2*

c) What term describes the aqueous copper sulphate at 30°C?

_____   *Hint 3*

**2** Consider the solubility curves of sodium chloride and potassium nitrate shown below.

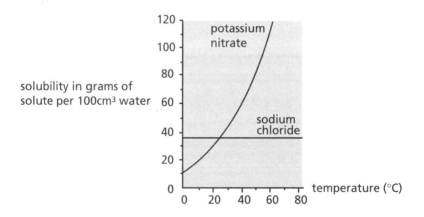

a) Which substance is the more soluble at 50°C? _____   *Hint 4*

b) At what temperature is the solubility of sodium chloride the same as the solubility of potassium nitrate?   *Hint 5*

_____

c) If 100 cm³ of potassium nitrate solution is cooled from 60°C to 30°C, what mass of solid will be deposited? *(Hint 6)*

_____

d) How much more potassium nitrate can be dissolved at 45°C than at 35°C? *(Hint 7)*

_____

**3** Three substances A, B and C, were added to water. Substances A and C dissolved whereas substance B only very slightly dissolved. Substances A and B are chlorides and substance C contained calcium.

Suggest identities for A, B and C (there is more than one possibility for each).

Substance A _____

Substance B _____ *(Hint 8)*

Substance C _____

**4** A student was asked to prepare some zinc sulphate (a soluble salt).

a) Name two reactants that could be used. *(Hint 9)*

_____

b) Which one should he use in excess? *(Hint 10)*

_____

c) How would you obtain the pure salt when the reaction was complete? *(Hint 11)*

_____

Another student was preparing the insoluble salt barium sulphate.

d) Name two reactants the student could use. *(Hint 12)*

_____

e) How would this student obtain a pure sample of this salt? 

_____ *(Hint 13)*

*Hints and answers follow*

# Salts – preparation and solubility

1. The solute is the substance that is to be dissolved.

2. The solvent is the liquid that dissolves the solid.

3. This is a solution in which no more solid will dissolve.

4. The lines on the graph tell you the maximum mass of solid that will dissolve at any given temperature.

5. Look for where these lines meet.

6. When the saturated solution is cooled the mass of salt that can be dissolved decreases. The excess is precipitated (dropped) out.

7. You need to calculate the difference between the masses that will dissolve to give you a saturated solution.

8. Quite simply you have got to learn the solubility of the salts as given in the table. See *improve your knowledge* section.

9. What methods are available for preparation of soluble salts?

10. The one that is easily removed once the reaction is finished should be in excess.

11. Learn this!

12. Barium sulphate formation is a characteristic test for the presence of the sulphate anion ($SO_4^{2-}$). How can it be made?

13. Learn this procedure!

## Answers

1 a) copper sulphate b) water c) saturated solution  2 a) potassium nitrate b) 23°C c) 80g d) 15g  3 substance A: any soluble chloride (i.e. any chloride except lead or silver)/substance B: any insoluble chloride (i.e. silver chloride or lead chloride)/substance C: calcium hydroxide / chloride / nitrate  4 a) zinc (or zinc carbonate or zinc hydroxide) + sulphuric acid b) the zinc (it is a solid metal) c) filter to remove the excess zinc then evaporate off the water (crystallisation) d) barium chloride solution + sulphuric acid e) filter to collect, rinse then evaporate off water

# Industrial processes 1

## Test your knowledge

**10 minutes**

**1.** Aluminium is high up in the _____ _____ and so is expected to be relatively reactive. However, it is less reactive than expected because upon exposure to air aluminium forms an _____ _____ which protects the surface. Aluminium is widely used in _____ _____ , where it is mixed with other metals to form an _____ which increases the _____ . The reactivity of aluminium can be further reduced by _____ .

**2.** Iron from the blast furnace has few uses because it is so _____ (i.e. breaks easily). Most iron is used in the form of _____ which is produced by converting the _____ in impure iron to _____ _____ by passing a high pressure of _____ through the molten iron.

**3.** Electrolysis of concentrated aqueous sodium chloride produces chlorine, _____ and sodium _____ . _____ turns damp pH paper red before _____ it. Some of the hydrogen produced is used to manufacture _____ which is used to make nitrogenous fertilisers. Domestic water is purified by adding _____ .

## Answers

**1** reactivity series / oxide coating / aircraft manufacture / alloy / strength / anodising **2** brittle / steel / carbon / carbon dioxide / oxygen **3** hydrogen / hydroxide / chlorine / bleaching / ammonia / chlorine

*If you got them all right, skip to page 64*

61

# Industrial processes 1

## Improve your knowledge

**1. Aluminium**

Aluminium is a reactive metal, yet aluminium window frames and greenhouse structures do not fizz and dissolve as soon as it starts to rain. This is because aluminium reacts with oxygen from the air to form a protective layer of aluminium oxide on the surface of the metal. This oxide coating can be artificially thickened by **anodising**.

The oxide layer is firstly removed by adding sodium hydroxide solution. The aluminium is then placed in dilute sulphuric acid, which is the electrolyte. The aluminium is made to be the anode (positive electrode). When a current flows, a **thicker** oxide layer is formed on the aluminium offering better protection.

The **strength** of aluminium can be increased by **alloying**. This involves mixing molten aluminium with other molten metals.

**2. Iron**

Iron from the blast furnace (containing many impurities) and pure iron have very few uses because they are so brittle, i.e. snap easily. Iron is mainly used as steel, which is made by lowering the carbon content of the iron extracted from the blast furnace.

Oxygen forced into the molten iron reacts with the carbon to form carbon dioxide, which, being a gas can easily escape, thus lowering the amount of carbon present in the iron. This produces **mild steel**.

Mild steel is used as the framework for car bodies, manufacture of 'tin' cans and making ship hulls. Steel can be mixed with other metals (i.e. alloying) to form 'alloy steels' with a variety of uses.

| Alloy steel | Property gained by alloy | Uses |
| --- | --- | --- |
| Stainless steel | Reduces rusting | Cutlery |
| Titanium steel | Increases springiness | Springs, machinery |
| Manganese steel | Increases strength | Metals for safes, steel in railway points |

## 3 Electrolysis of concentrated sodium chloride solution

Industrial electrolysis of concentrated aqueous sodium chloride solution produces hydrogen, chlorine, and sodium hydroxide.

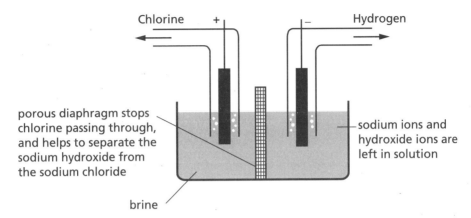

The electrolyte contains the following ions: $Na^+$ $H^+$ $Cl^-$ $OH^-$. During electrolysis the hydrogen cation ($H^+$) moves to the cathode where it picks up electrons to form hydrogen gas. Chlorine anions ($Cl^-$) move to the anode and lose electrons to form chlorine gas. The sodium cation ($Na^+$) and the hydroxide anion ($OH^-$) are left behind as sodium hydroxide. The chemical test and uses for each of the products are given in the table below:

| Chemical | Test | Uses |
| --- | --- | --- |
| Hydrogen | A lighted splint gives a 'squeaky pop' | Manufacture of ammonia |
| Chlorine | Turns pH paper red before bleaching it | Making bleach, water treatment |
| Sodium hydroxide | Turns pH paper blue | Making soap and paper |

## Check list

**Are you sure you understand the following key terms?**

**reactivity of aluminium / anodising / alloying / production of steel / electrolysis of aqueous sodium chloride**

*Now learn how to use this knowledge*

# Industrial processes 1

## Use your knowledge

20 minutes

**1** Aluminium is mixed with magnesium and copper to form the alloy 'duralumin' which is used in aircraft manufacture because it is strong and light. Anodised aluminium is used to make door/window frames, for example in greenhouses.

a) Define the term 'alloy'.

*Hint 1*

b) Why is an aluminium **alloy** used in aircraft manufacture rather than pure aluminium?

*Hint 2*

c) Why is aluminium washed with sodium hydroxide before it is anodised?

*Hint 3*

d) What advantage does anodised aluminium have over untreated aluminium?

*Hint 4*

During an investigation into the reactivity series a student noted that zinc reacted more vigorously than aluminium with acid. The student concluded that zinc is more reactive than aluminium.

e) Given that aluminium is actually more reactive than zinc, explain why the student's interpretation was incorrect.

*Hint 5*

**2** Oxygen is forced into molten iron (from the blast furnace) to convert non-metal impurities into acidic oxides. Calcium carbonate is then added leaving pure iron. Calculated quantities of carbon or metals are added to form alloys of iron.

a) Is calcium carbonate an acidic or basic substance?

*Hint 6*

b) Use evidence from the question to explain your answer to part a). *(Hint 7)*

c) What name is given to alloys of iron? *(Hint 8)*

d) Give the name of a metal alloy of iron and state one of its uses. *(Hint 9)*

**3** Concentrated aqueous sodium chloride solution is electrolysed to give three industrially important chemicals. Two of these products are gaseous and the third is a liquid that turns pH paper dark blue.

a) What are the three chemicals produced? *(Hint 10)*

b) One of the gases is a colourless gas that gives a 'squeaky pop' when exposed to a lighted splint. Which gas is this? *(Hint 11)*

c) Give one industrial use of this gas. *(Hint 12)*

d) Suggest a likely pH range for the liquid that turns pH paper dark blue. *(Hint 13)*

*Hints and answers follow*

# Industrial processes 1

*Hints*

1. Learn this definition.

2. What property of aluminium does alloying improve?

3. What is the first stage in the anodising of aluminium?

4. Anodised aluminium has a thicker oxide coating than untreated aluminium.

5. Why is aluminium less reactive than expected?

6. What does calcium carbonate react with?

7. See above.

8. Learn this!

9. Examples are iron mixed with manganese or titanium.

10. Learn this!

11. Learn this too!!

12. It is used to make a chemical that makes nitrogenous fertilisers.

13. A liquid that turns pH paper dark blue is a strong base.

*Answers*

1 a) a mixture of metals b) alloying aluminium increases the strength c) to remove the oxide coating formed naturally on aluminium d) the oxide layer is thicker, thus the aluminium is more protected e) the oxide layer formed naturally on aluminium shields the aluminium surface 2 a) basic b) it reacts with the acidic impurities c) steels d) any one of: stainless steel – cutlery; titanium steel – machinery / springs ; manganese steel – safes 3 a) hydrogen, chlorine, and sodium hydroxide b) hydrogen c) manufacture of ammonia d) 10-14

# Industrial processes 2

## Test your knowledge

**10 minutes**

**1** Ethanol is produced by _____ of _____ . Yeast acts as a biological _____ . The sugar reacts to produce ethanol and _____ _____ which is allowed to escape into the air. This method of producing ethanol produces a weak solution that can be made more concentrated by _____ _____ . The role of fermentation is also employed in making bread. Here, the gas is trapped in the dough causing it to _____ .

**2** Ethanol is used industrially as a _____ and is present in methylated spirit where it is mixed with _____ to make it unfit to drink. In countries where there is a limited supply of oil but good conditions for growing sugar, ethanol is produced and used as a _____ .

**3** Sulphuric acid is manufactured by allowing sulphur to burn in air forming _____ _____ which is further oxidised to _____ _____ before being dissolved in _____ to form sulphuric acid.

List two uses of sulphuric acid.

a) _____

b) _____

## Answers

**1** fermentation / carbohydrates (sugars) / catalyst / carbon dioxide / fractional distillation / rise **2** solvent / methanol / fuel **3** sulphur dioxide / sulphur trioxide / water / any two from: car batteries manufacture of fertilisers / detergents / chemicals / paints

*If you got them all right, skip to page 70*

# Industrial processes 2

## Improve your knowledge

**20 minutes**

### 1. Ethanol

Ethanol is the alcohol found in alcoholic drinks. We only need to be concerned with its biological production from yeast, i.e. **fermentation**. This is achieved by mixing water, sugar (a carbohydrate) and yeast (a biological catalyst) at about 40°C. The word equation for the process is:

carbohydrate → ethanol + carbon dioxide

This fermentation of sugars only produces ethanol concentrations of about 7% (at higher alcohol concentrations the yeast starts to die). However, some alcoholic drinks, e.g. gin, have ethanol concentrations of around 40%. Increased ethanol concentrations can be obtained by **fractional distillation.**

Fermentation also occurs during the baking of bread. This time it is the carbon dioxide that is of importance. Bubbles of gas become trapped in the dough and when placed in the oven the gas expands, hence the bread rises.

### 2. Uses and abuses of alcohol

The table below shows the social issues concerning alcoholic drinks.

| Advantages | Disadvantages |
| --- | --- |
| Alcohol industry employs thousands of people | Can be addictive which often leads to serious health problems e.g. liver failure |
| Drinking in moderation can be a good way to relax and socialise | Drunk drivers are the major cause of deaths on the road |
|  | It is a depressant drug |

Ethanol is the main constituent of methylated spirits; it is mixed with methanol (a highly toxic chemical) to make it unfit to drink. Ethanol is also used as an industrial solvent.

With the limited supplies of crude oil being rapidly used up, some countries are considering using ethanol as a fuel. Countries where this is happening usually have poor supplies of crude oil but good conditions for producing sugar, e.g. Brazil.

## 3 Sulphuric acid

Sulphuric acid ($H_2SO_4$) is manufactured in three stages.

a) **Combustion of sulphur**
$S + O_2 \rightarrow SO_2$

b) **Oxidation of the sulphur dioxide**
$2SO_2 + O_2 \rightarrow 2SO_3$ (This reaction is done using a vanadium (V) oxide catalyst at about 450°C.)

c) **Dissolving the sulphur trioxide in water**
$SO_3 + H_2O \rightarrow H_2SO_4$

Sulphuric acid is an industrially important chemical. The pie chart on the right represents the main uses of sulphuric acid.

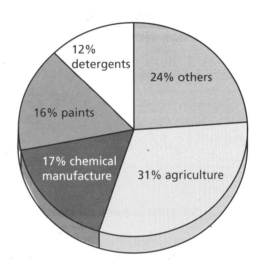

12% detergents
24% others
16% paints
17% chemical manufacture
31% agriculture

# Check list

**Are you sure you understand the following key terms?**

**formation of ethanol by fermentation / uses of ethanol / formation of sulphuric acid / uses of sulphuric acid**

*Now learn how to use this knowledge*

# Industrial processes 2

## Use your knowledge

**1** Wine can be made by adding sugar and yeast to fruit and leaving this mixture to settle for a long period of time. Whisky, however, has to be produced in factories called distilleries.

a) By what process is wine produced?

*Hint 1*

b) Why do you think that whisky cannot be produced by this process?

c) What chemical process do you think occurs in the distilleries?

**2** In the manufacture of sulphuric acid, sulphur (from sulphide ores) is burnt in air to produce sulphur dioxide, which is further mixed with oxygen and passed over vanadium oxide at 450°C. The product of this reaction is dissolved in water to form sulphuric acid ($H_2SO_4$).

a) Write a word equation for the reaction between sulphur and oxygen.

*Hint 4*

b) What is the function of the vanadium oxide?

c) What gas is produced when the mixture of gases is passed over the vanadium oxide?

*Hint 6*

A factory manufacturing sulphuric acid produces acid to supply a local agricultural company and also to a local carmaker for use in batteries.

d) What will the agricultural company use the sulphuric acid for? *(Hint 7)*

_____

3. Some estimates suggest that deposits of crude oil will run out in under 50 years' time. Some countries are preparing for this by developing alternative energy sources. Some hot countries, e.g. Brazil, are using ethanol as a fuel in specially developed cars. The ethanol is obtained from sugar. For some other countries, e.g. Britain, it is difficult to grow enough sugar.

a) Why is Brazil a good country for growing sugar cane? *(Hint 8)*

_____

b) Why is Britain unable to grow enough sugar? *(Hint 9)*

_____

Upon combustion of ethanol the products are carbon dioxide and water.

c) Explain why ethanol may be considered a more environmentally friendly fuel than petrol. *(Hint 10)*

_____

*Hints and answers follow*

# Industrial processes 2

## Hints

1. Wine is an alcoholic drink.

2. Whisky has an alcohol content of about 40%.

3. What method is used to produce such high alcohol concentrations?

4. The reactants are sulphur and oxygen, the product is sulphur dioxide (for more help in writing word equations see chapter 2).

5. It performs the same role as iron in the manufacture of ammonia.

6. It is the gas produced when sulphur dioxide is oxidised.

7. It is used to produce a product that promotes plant growth.

8. What conditions are needed for sugar to grow?

9. What are the climatic conditions of Britain compared with Brazil?

10. Combustion of petrol produces many harmful pollutants.

## Answers

**1** a) fermentation b) the alcohol content of whisky is too high for it to be produced by fermentation alone c) fractional distillation **2** a) sulphur + oxygen → sulphur dioxide b) catalyst c) sulphur trioxide d) manufacturing fertilisers **3** a) ideal climate i.e. warm and wet b) not an ideal climate, i.e. too wet and not warm enough c) the products of combustion of ethanol are not as environmentally damaging

# Products from ores

## Test your knowledge

**10 minutes**

**1** There are two methods of metal extraction available, the _____ _____ and _____ . Metals above zinc in the reactivity series (see chapter 5) are extracted by _____ . Zinc and other less reactive metals are extracted by the _____ _____ . Some less reactive metals, e.g. gold are found _____ i.e. no separation is required.

**2** Iron is extracted from its ore _____ ($Fe_2O_3$) using the _____ _____ . Here iron ore, _____ and _____ are added at the top and _____ is blasted in at the bottom. Carbon dioxide is formed which is _____ to carbon monoxide, which further reduces iron oxide to _____ . The purpose of the _____ in the blast furnace is to react with the silicon impurities present in iron ore to form _____ which is used on road surfaces and in _____ .

**3** During the extraction of aluminium from _____ (its ore – aluminium oxide) the ore is dissolved in molten _____ ; this lowers the melting point of aluminium oxide making the whole process _____ . A liquid that conducts electricity is called an _____ . When dissolved, the aluminium oxide exists as $Al^{3+}$ cations which move to the _____ (negative electrode), and $O^{2-}$ anions which move to the _____ (positive electrode).

## Answers

**1** blast furnace / electrolysis / electrolysis / blast furnace / pure  **2** haematite / blast furnace / limestone (calcium carbonate) and coke / (hot) air / reduced / iron / limestone/calcium carbonate / slag / making cement  **3** bauxite / cryolite / cheaper / electrolyte / cathode / anode

If you got them all right, skip to page 76

# Products from ores

## Improve your knowledge — 20 minutes

### 1. Ores and metal extraction

An ore is the form in which some metals are found in the Earth's crust. Ores are usually metal oxides or substances that can easily be changed into the metal oxide. Some less reactive metals, e.g. gold, are found pure, i.e. no separation is required.

There are two methods of separating metals from their ores.

a) The blast furnace - for zinc and other less reactive metals

b) Electrolysis - for metals above zinc in the reactivity series (see chapter 5)

Both methods of extraction involve **reducing** the metal oxide to the metal. Reduction is the loss of oxygen (reduction is the opposite of oxidation).

### 2. The blast furnace

Iron is extracted from its ore haematite ($Fe_2O_3$) using the blast furnace.

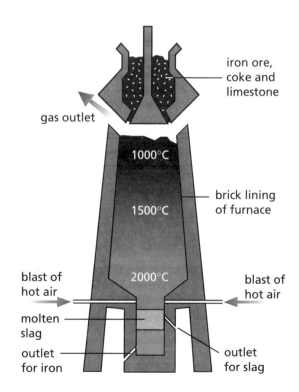

There are **five** important reactions that occur in the blast furnace.

1) The coke reacts with oxygen from the air.

   $C + O_2 \rightarrow CO_2$

2) This carbon dioxide is reduced to carbon monoxide.

   $CO_2 + C \rightarrow 2CO$

3) Carbon monoxide reduces the iron oxide:

$$3CO + Fe_2O_3 \rightarrow 2Fe + 3CO_2$$

4) Limestone decomposes giving carbon dioxide and calcium oxide:

$$CaCO_3 \rightarrow CaO + CO_2$$

5) The calcium oxide reacts with $SiO_2$ (the main impurity in haematite) to produce slag:

$$CaO + SiO_2 \rightarrow CaSiO_3$$

The slag is used as a fertiliser and on road surfaces.

## 3 Electrolysis

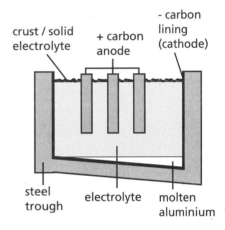

This is a more powerful method of reduction but is more expensive because a large amount of electricity is used. Aluminium is extracted from its ore, bauxite ($Al_2O_3$), by electrolysis.

The purified bauxite is dissolved in cryolite (to lower the melting point thus reducing costs). This liquid conducts electricity so is called an **electrolyte**. In this solution the aluminium oxide exists as aluminium cations ($Al^{3+}$) and oxygen anions ($O^{2-}$). The positively charged **cations** move to the **cathode** where they pick up electrons and aluminium metal is formed.

The negatively charged **anions** move to the **anode** where they give up electrons to form oxygen gas. This immediately reacts with the carbon anodes (because of the high temperatures present) forming carbon dioxide gas.

# Check list

Are you sure you understand the following key terms?

**extraction of reactive metals e.g. Al / extraction of less reactive metals e.g. Fe / oxidation and reduction**

*Now learn how to use this knowledge*

# Products from ores

## Use your knowledge

20 minutes

**1** Zinc is largely found in the earth's crust as the ore *zinc blende* (ZnS). Silver however, is found pure, i.e. as the unreacted element. Before the zinc can be extracted it has to undergo chemical treatment.

a) What must the zinc ore be converted to in order for the zinc to be extracted?  *(Hint 1)*

_____

b) In order to separate the zinc from the treated ore it is heated with coke (carbon). Carbon dioxide is one of the products of this treatment. In what way is the function of the coke similar to the function of carbon monoxide in the blast furnace?  *(Hint 2)*

_____

c) Why do you think that electrolysis is not used to extract zinc given that it is quicker than the above method?  *(Hint 3)*

_____

d) What explanation can you offer for the silver being found pure but the zinc being found as an ore?  *(Hint 4)*

_____

**2** Aluminium and iron are extracted from their ores by two different methods.

a) What are the two methods of extracting metals?  *(Hint 5)*

_____

b) During the extraction of aluminium, the ore is first dissolved in molten cryolite. Cryolite is expensive so why do you think it is used?  *(Hint 6)*

_____

c) Why do you think that we are constantly being encouraged to recycle aluminium more than any other metal even though aluminium is the most abundant metal in the Earth's crust?

*Hint 7*

_____

**3** During the extraction of iron, coke (carbon), limestone (calcium carbonate) and haematite (iron ore) are added to the blast furnace. At about 1600°C the haematite is reduced to iron.

a) What do you understand by the term reduced?

*Hint 8*

_____

b) What are the two functions of the limestone?

*Hint 9*

_____

_____

_____

c) Give two uses for the solid waste product from the blast furnace.

*Hint 10*

_____

_____

d) In what way does the following reaction reduce the costs associated with running a blast furnace?

*Hint 11*

$C + O_2 \rightarrow CO_2$

_____

*Hints and answers follow*

# Products from ores

## Hints

1. What is the chemical name for the ores that aluminium and iron are extracted from and how do these differ from the zinc ore?

2. The carbon is **oxidised** to carbon dioxide.

3. What is the major disadvantage with electrolysis as a method of extraction?

4. Which types of metals are found unreacted in the Earth's crust?

5. Learn this!

6. How does the melting point of pure bauxite compare with that of bauxite in cryolite?

7. How do you think the costs of extraction compare with the costs of recycling?

8. Learn this definition.

9. One of the products is further reduced and the other reacts with an impurity.

10. The waste product is called 'slag' and you need to learn the uses of slag.

11. This reaction is 'exothermic', i.e. gives out heat.

## Answers

1 a) the ore must be converted to the oxide b) both **reduce** the ore c) electrolysis is too expensive d) unreactive metals exist in their pure form; zinc is more reactive than silver 2 a) blast furnace and electrolysis b) the melting point of aluminium oxide is lower when it is mixed with the cryolite thus energy costs are lower c) extraction of aluminium is more expensive than recycling because of the large amount of electricity needed 3 a) loss of oxygen b) when limestone is heated carbon dioxide and calcium oxide are produced; the carbon dioxide is reduced to carbon monoxide which reduces the iron oxide; the calcium oxide reacts with silicon oxide (impurity in haematite) to form slag (calcium silicate) c) fertiliser, road surfaces d) the reaction is exothermic (gives out heat), this reduces the amount of energy needed to generate the temperatures of up to 2000°C

# Products from oil

## Test your knowledge

**1** Crude oil is formed in the Earth's _____ by the long-term effects of _____ and _____ on the remains of organisms (marine deposits). Crude oil deposits are formed in _____ rock below _____ rock. Crude oil is a _____ of different _____ . These are substances that contain _____ and _____ only. Often, where oil is formed, _____ _____ is also formed. As they both are _____ dense than water, they rise up and collect below the _____ rock, and can be collected by _____ through this rock.

**2** Crude oil can be separated by _____ _____ . This is done by _____ the oil and passing it through a _____ _____ . Hydrocarbons with a small number of carbon atoms in the molecule will condense at a _____ temperature. Larger molecules have a _____ boiling point. They are also _____ (i.e. don't flow very easily). Hydrocarbons that are highly flammable are ones with _____ molecules.

**3** Burning hydrocarbons in excess oxygen (air) produces heat, _____ _____ and _____ . The gaseous product turns limewater _____ . With a lack of oxygen, incomplete combustion occurs which produces _____ _____ and/or _____ as a product.

**4** Large hydrocarbons are broken down into smaller, more useful products by _____ . Some of these products are used as _____ , others are joined together to make very large molecules called _____ .

**Answers**

**1** crust / heat and pressure / porous / non-porous / mixture / hydrocarbons / carbon and hydrogen / natural gas or methane / less / non-porous / drilling **2** fractional distillation / evaporating / fractionating column / low / high / viscous / small **3** carbon dioxide / water / milky (cloudy) / carbon monoxide / carbon **4** cracking / fuels / polymers

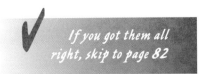

If you got them all right, skip to page 82

# Products from oil
## Improve your knowledge

**1. Formation**

Crude oil is formed in porous rock by the long-term effects of heat and pressure on marine deposits (dead animals/plants). Where oil is found, natural gas (methane) is also usually found. Oil and natural gas, being less dense than water, rise up through the porous rock and collect under non-porous rock.

**2. Separation**

The crude oil obtained by drilling through the non-porous rock is simply a mixture of lots of different **hydrocarbons**. Hydrocarbons are substances that contain hydrogen and carbon **only**. As with any mixture of liquids, the crude oil is separated by **fractional distillation**. Before separation the crude oil is heated until it evaporates.

The table below represents the properties of each of the different fractions obtained.

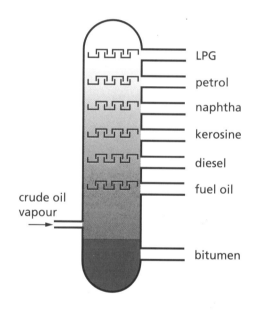

| Name of fraction | Number of carbon atoms | Boiling range °C | Uses |
|---|---|---|---|
| Light Petrol Gases | 1-4 | <25 | Portable fuels e.g. camping gas |
| Petrol | 4-12 | 25-60 | Car fuel |
| Naphtha | 7-14 | 60-180 | Making other chemicals e.g. fertilisers |

| Kerosine | 9-16 | 180-220 | Home heating (paraffin) |
| Diesel | 15-25 | 220-250 | Diesel car engines |
| Fuel oil | 20-70 | 250-330 | School central heating |
| Bitumen | >70 | >350 | Road surfaces |

## 3 Fuels

Many hydrocarbons are used as fuels, particularly methane ($CH_4$), ethane ($C_2H_6$), propane ($C_3H_8$) and butane ($C_4H_{10}$). As well as providing energy (i.e. heat), combustion (burning) of fuels in a plentiful supply of air also gives carbon dioxide and water. If carbon dioxide is bubbled through limewater the limewater turns 'milky'. In a limited supply of oxygen, carbon monoxide is produced instead of carbon dioxide. Carbon monoxide is a colourless, odourless and highly toxic gas.

## 4 Cracking and polymerisation

Cracking is the process by which long chain hydrocarbons are broken down into smaller (more useful) hydrocarbons. Cracking is achieved by passing the hydrocarbon vapour over an aluminium oxide catalyst at 400°C. Some of the small hydrocarbons formed during cracking are used as fuels, others undergo polymerisation. Polymerisation is the joining together of lots of small molecules to make one big one.

Examples and uses of polymers are given in the table below.

| Polymer | Properties | Uses |
| --- | --- | --- |
| Polythene | Light, flexible, strong | Carrier bags |
| Poly(propene) | Strong, rigid, light | Plastic chairs |
| Poly(styrene) | Light, insulator of heat | House insulation, packaging of electrical goods |

## Check list

Are you sure you understand the following key terms?

formation of oil / fractional distillation / cracking / polymerisation / properties and uses of polymers / formation of pollutants

*Now learn how to use this knowledge*

# Products from oil

## Use your knowledge

20 minutes

1. The table below shows some of the fractions of oil and the boiling range of the constituent hydrocarbons.

| Fraction | Boiling range °C |
|---|---|
| LPG | < 25 |
| Petrol |  |
| Naphtha | 60–180 |
| Diesel | 220–250 |

a) Suggest a boiling range for petrol _____ . *(Hint 1)*

b) Using information in the table, what can you determine about the size of the molecules of petrol compared with the molecules of naphtha? *(Hint 2)*

_____

c) Explain why petrol is considered to be more of a fire hazard than diesel is. *(Hint 3)*

_____
_____

In an experiment, one of the hydrocarbons in the naphtha fraction was broken down to give one of the hydrocarbons present in the petrol fraction as well as another small molecule.

d) What is this process called and how is this carried out industrially? *(Hint 4)*

_____

The other small molecule is joined together with lots of other molecules to form polythene.

e) What name is given to this process? *(Hint 5)*

_____

f) What properties does polythene possess that makes it suitable to be used for plastic carrier bags? *(Hint 6)*

_____

**2** The table below gives the names and formulae of some hydrocarbons. Some of the information is missing.

| Name of hydrocarbon | Formula |
|---|---|
| Methane | |
| | $C_2H_6$ |
| Propane | |
| | $C_4H_{10}$ |

a) Complete the table. *(Hint 7)*

b) Which one of the above has the highest boiling point? *(Hint 8)*

_____

c) Which one of the above is known as 'natural gas'? *(Hint 9)*

_____

d) What are the products of the complete combustion of hydrocarbons? *(Hint 10)*

_____

*Hints and answers follow*

# Products from oil

### Hints

1. The petrol fraction is in between LPG and naphtha, so where will the boiling range be?

2. What is the connection between the size of a molecule and the boiling point?

3. Which fuel is more likely to turn into a vapour?

4. This is the process where long chain hydrocarbons are broken down into smaller (more useful) hydrocarbons.

5. Here lots of small molecules are joining together to make a large molecule.

6. Properties are things like strength, density, how easily it can be made into different shapes.

7. All of these hydrocarbons are found in the LPG fraction.

8. What is the connection between molecular size and boiling point?

9. It is the smallest sized hydrocarbon.

10. Learn these!

### Answers

1 a) 25-60°C b) petrol molecules must be smaller than diesel (because petrol has a lower boiling point) c) petrol has smaller molecules than diesel hence petrol will turn into a vapour more easily (it is the vapours that are the fire hazard) d) cracking, passing vapour at 400°C (always put units in where appropriate) over an aluminium oxide catalyst e) polymerisation f) light, flexible, strong and easily made into shape 2 a) reading left to right... $CH_4$, ethane, $C_3H_8$, butane b) butane (it has the largest size molecule of the four hydrocarbons) c) methane d) carbon dioxide, water and heat

# Rocks and plate tectonics

## Test your knowledge

**1** There are three types of rock: igneous, _____ and metamorphic.

**2** Igneous rocks form when _____ cools and solidifies.

**3** Sedimentary rocks can be recognised because they are built up in _____ . The sediments are compressed or _____ to form sedimentary rock.

**4** Metamorphic rocks are formed from igneous and sedimentary rocks which are changed by _____ and/or _____ .

**5** Only _____ rocks contain fossils. This is because fossils are destroyed by the heat and pressure when _____ and _____ rocks form.

**6** The breakdown of rock by air and water is called _____ .

**7** The Earth's crust is split into huge sections called _____ , which can move. This is known as the theory of _____ _____ .

**8** Volcanoes are often found at the _____ of plates which are moving. When plates collide one may sink below the other and _____ .

**Answers**

1 sedimentary  2 magma  3 layers/strata, consolidated  4 heat / pressure  5 sedimentary / igneous / metamorphic  6 weathering  7 plates / plate tectonics  8 edges / melt

If you got them all right, skip to page 88

# Rocks and plate tectonics

## Improve your knowledge

**1** There are three rock types, **igneous, sedimentary** and **metamorphic**, which are classified according to how they are formed.

**2** Igneous rocks form when **magma** (molten rock) cools and solidifies.

**3** Sedimentary rocks can be formed from fragments of igneous or metamorphic rocks or from fragments of shells which build up in layers over millions of years. Eventually the great pressure turns the sediments into rock, i.e. **consolidates** them. The deepest sediments were buried first, i.e. they are the oldest.

**4** Metamorphic rocks are formed from igneous and sedimentary rocks which are changed by heat and/or pressure.

**5** Sedimentary rocks often contain **fossils** – but fossils are not found in igneous or metamorphic rocks – they are destroyed by the heat and pressure when these rocks are formed.

**6** All rocks are slowly broken down by air and water (**weathering**). For example, a rock may be dissolved by acid rain or it may be slowly worn away by being rolled along in a fast-flowing stream. Weathering produces tiny particles of rock which are washed away and end up as sediments.

The relationship between igneous, sedimentary and metamorphic rocks is shown by the rock cycle overleaf.

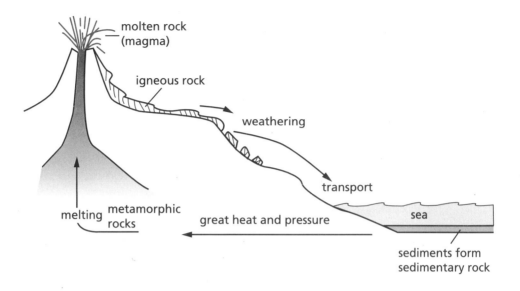

**7** The Earth's crust is split into gigantic sections called **plates**. These float on the molten material below and can move apart or collide. This is the theory of **plate tectonics**.

**8** When two plates collide, one of them may be pushed under the other and melt. The rock is therefore recycled back into magma. If one plate rides up over the edge of another plate or if neither plate sinks, the rocks are pushed upward and can form huge mountains, e.g. the Himalayas. Volcanoes and earthquakes are also common at the edge of tectonic plates. If two plates move apart, magma may be released violently in a volcanic eruption – it cools to form igneous rocks.

## Check list

**Are you sure you understand the following key ideas?**

**formation of the three types of rock / weathering / plate tectonics / volcanoes and earthquakes**

*Now learn how to use this knowledge*

# Rocks and plate tectonics

*Use your knowledge*

The diagram shows the movement of 3 tectonic plates and the position of the South American and African continents.

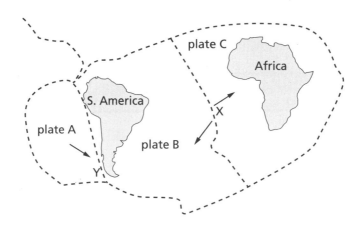

**1** What would you expect to happen at X?  *(Hint 1,2)*

**2** What type of rocks would be expected to form at X?  *(Hint 3)*

**3** Plate A is colliding with and being forced below plate B. Why will metamorphic rock form below area Y?  *(Hint 4)*

**4** Why would you not expect to find fossils in the rocks below Y?  *(Hint 5)*

**5** How can an igneous rock on a mountain in Scotland end up as a sedimentary rock below the North Sea?  *(Hint 6,7,8)*

*Hints and answers follow*

# Rocks and plate tectonics

1. In what direction are the two plates moving?

2. What features do you often find when plates move like this?

3. Molten material will spurt up from below the sea. What kind of rocks form when this occurs?

4. What will happen to plate A when it is forced under plate B?

5. What conditions are needed for metamorphic rocks to form?

6. What happens to all rocks which are above the ground level?

7. How can rock particles be transported from one area to another?

8. How are sediments or particles turned into rock?

## Answers

**1** release of magma / volcanic activity **2** there are conditions of great heat and pressure **3** igneous **4** they will be destroyed by the heat and pressure, the particles may be washed into rivers by the rain, the rivers carry the particles to the seas where they build up as sediments **5** the rock will be weathered, the particles may

# Mock exam

1 The element nitrogen, N, is in group 5 of the periodic table. It exists as a diatomic molecule and forms many different compounds.

   a) How many protons are there in the nucleus of a nitrogen atom?
   _____ (1)

   b) How many neutrons are there in the nucleus of a nitrogen atom?
   _____ (1)

   c) Write down the electronic structure of a nitrogen atom.
   _____ (1)

   d) Write down the formula of a nitrogen molecule.
   _____ (1)

   e) Write down the formula and the electronic structure of the ion that you would expect nitrogen to form.
   _____ (2)

   [6]

2 Nitrogen reacts with hydrogen to form a compound called ammonia which contains only nitrogen and hydrogen. Ammonia is made industrially by heating the gases up to 300°C over an iron oxide catalyst and by applying high pressures of around 250 atmospheres.

   a) Complete and balance the equation for the formation of ammonia.

   _____ $N_2$ + _____ $H_2$ → _____ (2)

   b) What is a catalyst?
   _____
   _____ (2)

c) If a catalyst is not used, then higher temperatures could be used instead. Explain how higher temperatures would have the same effect as a catalyst.

_____

_____

_____ (3)

d) Why do you think using high temperatures would make the ammonia more expensive?

_____

_____ (2)

[9]

3  Ammonia is used to make nitric acid, $HNO_3$, by heating it with oxygen to form first nitrogen monoxide, NO, and then nitrogen dioxide, $NO_2$. Nitrogen dioxide is then reacted with water and more oxygen to form nitric acid.

a) Write a balanced equation for the formation of $NO_2$ from NO and $O_2$.

_____ (1)

b) What type of reaction is this? _____ (1)

c) Nitric acid is often used to make fertilisers such as ammonium nitrate, $NH_4NO_3$, which is particularly effective because of its high nitrogen content. Why do fertilisers need to contain nitrogen to be effective?

_____

_____ (2)

[4]

4  The element chromium is a metal and has the symbol Cr. Like most metals it has a high melting and boiling point and is physically strong. It reacts with fluorine gas, a halogen, to form chromium fluoride, $CrF_3$.

a) State two other **physical** properties you would expect chromium to show as a typical metal.

_____

_____ (2)

b) What type of bonding is present in chromium fluoride?

_____ (1)

c) What is the valency of chromium in chromium fluoride?

_____ (1)

d) What would you expect the formula of chromium oxide to be?

_____ (1)

[5]

**5** Small amounts of chromium and nickel are mixed with iron to produce stainless steel. This is often used to make things which need to be strong but which are exposed to water and air such as cutlery and medical instruments.

a) What is the name for a mixture of metals such as steel?

_____ (1)

b) Give another used for stainless steel.

_____ (1)

c) Pure iron is cheaper than stainless steel. Why is it not suitable for manufacturing objects which are exposed to the air and water?

_____
_____ (1)

d) Chromium is a transition metal and its compounds are used to manufacture paint. Why is this so? _____

_____ (1)

[4]

**6** The bar chart below shows the approximate dates by which the main ores of some metals will have been used up at the present rate of use.

a) Which metal will be used up last? _____ (1)

b) Which metal should show the greatest price increase over the next few years? Give a reason for your answer.

_____ (2)

c) Iron is more abundant in the Earth's crust than chromium, yet it is likely to run out first. Why do you think this is?

_____
_____ (2)
[5]

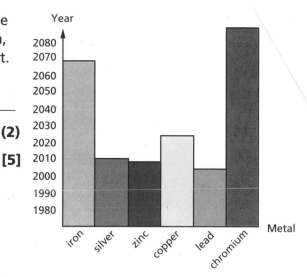

**7** Iron is extracted from its main ore ($Fe_2O_3$) using the blast furnace.

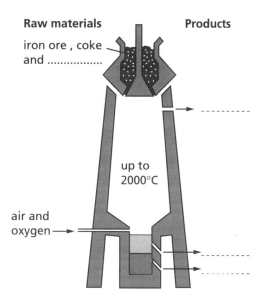

Raw materials: iron ore, coke and ..................

air and oxygen →

up to 2000°C

Products

a) On the diagram, label the one missing raw material and the three missing products. **(4)**

Some metals such as aluminium and sodium cannot be extracted from their ores by the blast furnace. They need to be extracted by a more powerful method of reduction.

b) What method of extraction is used for these metals?

_____ (1)

c) Why do these metals have to be extracted by a stronger method of reduction? _____ (2)

Inside the blast furnace the iron ore is reduced to iron by carbon monoxide. The equation below represents how carbon monoxide is formed.

$$CO_2 \text{ (g)} + C \text{ (s)} \rightarrow 2CO \text{ (g)}$$

d) What is reduction? _____ (1)

e) In the above equation which substance has been oxidised?
_____ (1)

[9]

**8** a) Suggest one reason why steel making plants are often situated very close to iron works. (1)
_____

b) Give two advantages and two disadvantages of building an iron works and a steel works close to a major town. (4)
_____
_____
_____

c) What is the major disadvantage of making seaside piers out of iron?
_____ (1)

[6]

**Total marks for paper = 48**

## Answers

**1** a) 7 b) 7 c) 2,5 d) $N_2$ e) $N^{3-}$. 2,8 **2** a) $N_2 + 3H_2 \rightarrow 2NH_3$ (1 mark for formula of ammonia) b) a substance which speeds up a chemical reaction (1) without being used up itself (1) c) at higher temperature the reacting particles/molecules have more energy OR speed (1) so that there are more successful collisions (1) which speeds up the rate of reaction(1) d) higher temperatures require more fuel (1) which costs more money raising the price of the ammonia (1) **3** a) $2NO + O_2 \rightarrow 2NO_2$ b) combustion or oxidation c) nitrogen is converted into nitrates in the soil (1) which are essential for plant growth (1) **4** a) any two from: good conductor of heat/good conductor of electricity/shiny when cut/malleable or easily bent into shape b) ionic c) 3 d) $Cr_2O_3$ **5** a) alloy b) car accessories/engines/ chemical plants/garden tools etc. c) iron reacts with oxygen in the presence of water (1) resulting in corrosion or rusting (1) causing the metal to become damaged/full of holes/break etc (1) d) because it is coloured **6** a) chromium b) lead (1), because it is the rarest of the metals shown (1) c) iron (1) is more widely used (1) **7** a) missing raw material: limestone ($CaCO_3$) (1), missing products: top – waste gases (mainly $CO_2$) (1), middle – slag (waste product $CaSiO_3$ calcium silicate) (1) bottom – molten iron (1) b) electrolysis (1) c) more reactive (1) and so are more difficult to separate from their ore (1) d) loss of oxygen (1) e) carbon (C) (1) **8** a) steel is manufactured from molten iron (1) b) advantages: employment (1), transport networks (1), disadvantages: any two from, air pollution (1), noise (1), increased traffic (1) c) the iron corrodes (rusts) particularly in sea (salty) water (1)

94